Medical and Biologic Effects of Environmental Pollutants

COPPER

Committee on
Medical and Biologic Effects of
Environmental Pollutants

DIVISION OF MEDICAL SCIENCES
ASSEMBLY OF LIFE SCIENCES
NATIONAL RESEARCH COUNCIL

NATIONAL ACADEMY
OF SCIENCES
WASHINGTON, D.C.   1977

Other volumes in the Medical and Biologic Effects of Environmental Pollutants series (formerly named Biologic Effects of Atmospheric Pollutants):

ASBESTOS (ISBN 0-309-01927-3)
CHROMIUM (ISBN 0-309-02217-7)
FLUORIDES (ISBN 0-309-01922-2)
PARTICULATE POLYCYCLIC ORGANIC MATTER (ISBN 0-309-02027-1)
LEAD (ISBN 0-309-01941-9)
MANGANESE (ISBN 0-309-02143-X)
VANADIUM (ISBN 0-309-02218-5)
NICKEL (ISBN 0-309-02314-9)
VAPOR-PHASE ORGANIC POLLUTANTS (ISBN 0-309-02441-2)
SELENIUM (ISBN 0-309-02503-6)
CHLORINE AND HYDROGEN CHLORIDE (ISBN 0-309-02519-2)

NOTICE: The project that is the subject of this report was approved by the Governing Board of the National Research Council, whose members are drawn from the Councils of the National Academy of Sciences, the National Academy of Engineering, and the Institute of Medicine. The members of the Committee responsible for the report were chosen for their special competences and with regard for appropriate balance.

This report has been reviewed by a group other than the authors according to procedures approved by a Report Review Committee consisting of members of the National Academy of Sciences, the National Academy of Engineering, and the Institute of Medicine.

The work on which this publication is based was performed pursuant to Contract No. 68-02-1226 with the Environmental Protection Agency.

Library of Congress Catalog Card Number 76-57888

International Standard Book Number 0-309-02536-2

Available from
Printing and Publishing Office
National Academy of Sciences
2101 Constitution Avenue
Washington, D.C. 20418

Printed in the United States of America

SUBCOMMITTEE ON COPPER

I. HERBERT SCHEINBERG, Department of Medicine, Albert Einstein College of Medicine, Bronx, New York, *Chairman*

WILLIAM B. BUCK, College of Veterinary Medicine, Iowa State University, Ames, Iowa

GEORGE E. CARTWRIGHT, Department of Medicine, University of Utah Medical Center, Salt Lake City, Utah

GEORGE K. DAVIS, Nutrition Laboratory, University of Florida, Gainesville, Florida

CHARLES R. DAWSON, Department of Chemistry, Columbia University, New York, New York

*JEAN M. MORGAN, Department of Medicine, University of Alabama School of Medicine, Birmingham, Alabama

KENNETH W. NELSON, American Smelting and Refining Company, New York, New York

CARL A. PRICE, Department of Biochemistry and Microbiology, Rutgers University, New Brunswick, New Jersey

IRMIN STERNLIEB, Department of Medicine, Albert Einstein College of Medicine, Bronx, New York

T. D. BOAZ, JR., Division of Medical Sciences, National Research Council, Washington, D.C., *Staff Officer*

*Deceased

COMMITTEE ON MEDICAL AND BIOLOGIC EFFECTS OF ENVIRONMENTAL POLLUTANTS

ERRATA

1. Pages 11–12, Footnote on page 12 should be on page 11

2. Page 17, line 2, "Bennets" should be "Bennetts"

3. Page 56, line 36, "TLVS" should be "TLV's"

4. On page 61, the paragraph under the heading, **Copper in Drinking Water**, should read as follows:

    Despite the widespread use of copper and brass plumbing, copper concentrations in drinking water rarely exceed the accepted standard of 1 ppm, unless water solutions of low pH are allowed to stand for a long time in such plumbing.

5. Recommendation 7, on page 64, should read as follows:

    Because copper is absorbed by the lungs, skin, and uterus, as well as by the gastrointestinal tract, a nationwide clinical investigation should be carried out to determine whether any long-term hazard of human copper toxicosis is possible from the added burden of body copper introduced parenterally through chronic hemodialysis, inhalation, or absorption from the skin, or from copper-containing intrauterine contraceptive devices.

6. Page 65, line 9, "particulate" should be "particulates"

7. Page 75, line 6, "AS_2O_3" should be "As_2O_3"

*Medical and Biologic Effects of
Environmental Pollutants: COPPER*
ISBN 0-309-02536-2

Acknowledgments

This document is a cooperative effort on the part of the members of the Subcommittee on Copper. Each chapter, and sometimes a separate section within a chapter, was the responsibility of a person expert in its subject matter. The summaries, conclusions, and recommendations are the consensus of the Subcommittee members; the entire document has been reviewed and approved by the Subcommittee.

Dr. I. Herbert Scheinberg, Chairman of the Subcommittee, wrote the introduction and was jointly responsible with Dr. Irmin Sternlieb for most of Chapter 5, which deals with human copper metabolism. Part of this chapter was prepared by Dr. Jean M. Morgan, who was unable to continue because of illness; she died before the report was completed. Dr. George K. Davis wrote Chapter 2, on copper in the ecosystem, and collaborated with Dr. William B. Buck on those sections of Chapter 4 that deal with copper toxicosis and its pathologic physiology. Dr. Buck had the primary responsibility for Chapter 4. Chapter 3, on copper in plants, was prepared by Dr. Charles R. Dawson and Dr. Carl A. Price. Dr. Dawson's material was prepared in collaboration with one of his colleagues, Dr. Kenneth G. Krul, not a member of the Subcommittee. Chapter 6, on copper as an industrial health hazard, was the work of Mr. Kenneth W. Nelson. Dr. George E. Cartwright reviewed the draft report, especially Chapter 5, and provided helpful comments and suggestions.

The Subcommittee is indebted to Ms. Avis Berman for her diligent editing of the report, and to Ms. Joan V. Stokes for locating and verifying the references. In obtaining information and references for the use of the Subcommittee in its deliberations and in the preparation of this report, the resources of the following were called upon: the National Research Council's Advisory Center on Toxicology, the National Academy of Sciences Library, the National Library of Medicine, the National Agricultural Library, the Library of Congress, and the Air Pollution Technical Information Center.

The Subcommittee acknowledges the assistance of the Environmental Studies Board, the National Academy of Sciences–National Academy of Engineering, and of divisions and offices of the National Research Council.

Contents

1

Introduction

Copper is essential to human life and health, and like all heavy metals, it is potentially toxic also. In response to this duality, biochemical mechanisms have evolved that control the absorption and excretion of copper. These mechanisms operate to offset the effects of temporary deficiency or excess of the metal in the diet.

Of the copper retained in the body, almost all plays a particular physiologic role as the prosthetic element of more than a dozen specific copper proteins, such as cytochrome c oxidase and tyrosinase. Thus only extremely small concentrations of free copper ions are normally found in body fluids. Because it is common for the toxicity of any heavy metal cation to be sharply diminished when it is bound to proteins or other macromolecules, the existence of these copper proteins—as well as of the homeostatic mechanisms governing absorption and excretion—make toxicosis from dietary copper extremely rare in man.

Only inordinately large amounts or concentrations of orally ingested copper are toxic. For example, acidic foods or beverages that have been in prolonged contact with copper metal may cause acute gastrointestinal disturbance. However, when copper enters the body by a parenteral route (for example, following inhalation, or absorption from burned skin or a contraceptive device in the uterine cavity), there is a significant possibility that toxicosis may result from amounts of copper that are innocuous when eaten.

1

2

Copper in the Ecosystem

Copper metal and its compounds have been used by man since pre-historic times and thus have been a part of the environment and eco-system in varying concentrations. An appreciation of the amount of copper being removed from sites of its natural source and released into the ecosystem in the United States and worldwide may be gained from the following statistics on copper production. From 1955 to 1958, annual U.S. production of recoverable copper was about 900,000 metric tons,[116] but by 1975 the production had risen to 1,260,000 metric tons.[574] Currently available statistics for world production of copper indicate that the amount of copper entering the ecosystem annually has steadily increased and now amounts to about 1,800,000 metric tons per year, especially in the industrially developed nations. The world trade in refined copper has been reported as amounting to 2,271,150 metric tons in 1973.[99]

Increased knowledge of the essential nature of copper in the metabolism of plants and animals, and of its versatility in many industrial and agricultural operations, has led to recognition of the widespread natural and man-made distribution of this element in concentrations ranging from the severely deficient to the toxic.

Copper is an element with atomic number 29, and an atomic weight of 63.546. It consists of two natural isotopes: copper-63 and -65, which constitute 69.09% and 30.91%, respectively, of the whole. It occurs in nature as the metal in the $+1$ and $+2$ valence states with ionic radii of 0.96Å and 0.72Å (for sixfold coordination), respectively. In the $+2$ state, it is isomorphous with Zn^{+2} (ionic radius, 0.74Å), Mg^{+2} (ionic radius, 0.66Å), and Fe^{+2} (ionic radius, 0.74Å).[610] Copper tends to occur in sulfide deposits, particularly in igneous rocks.

The concentration of copper in the continental crust—generally estimated at 50 ppm—[438] tends to be highest in the ferromagnesium minerals, such as the basalts pyroxene and biotite, where it averages 140 ppm. Copper commonly forms organic complexes, although coal is relatively low in copper. Sandstones contain 10–40 ppm, shales, 30–150 ppm, and marine black shales, 20–300 ppm.

In the sedimentary cycle, copper is concentrated in the clay mineral fractions with a slight inclination toward enrichment in those clays rich in organic carbon, and it is notably concentrated in sedimentary manganese oxides with values up to one-tenth of a percent. Recovery of copper and other metals from ocean and lake floors may prove more profitable than the recovery of the manganese nodules for which such commercial ventures were originally designed. More than 99.9% of the copper carried to the ocean is precipitated, mostly with the clays and partially with manganese oxides. It is probable that much of the copper reported in surface waters comes from contamination with industrial metallurgic waste.

Factors influencing the relationship between copper in the parent rock and the derivative soil include the degree of weathering, the nature and intensity of the soil formation, drainage, pH, oxidation–reduction potential, and the amount of organic matter in the soil.[14] Since copper in rocks is likely to be more mobile under acid than alkaline conditions, the relation of pH to copper mobility in the environment has been of concern to agriculturalists and biologists. Alkaline conditions in the soil and surface waters favor precipitation of copper, which may cause deficiencies in plants and minimize the effectiveness of copper-containing molluscicides. Conversely, acid conditions promote solubility of copper, increase the concentration of ionic copper, and thereby change the microorganism and other aquatic animal populations, depending upon tolerance for various levels of copper in solution.[59,143,302] The report of acid rain occurring in various parts of the world is pertinent to this consideration.[312]

The amount of clay in the soil is a key factor in its capacity for cation exchange and consequent movement or retention of copper. The moisture content of the soil, a key factor in microorganism activity, also influences the availability of copper. Because of these factors, the concentration of copper in the soil is subject to considerable variation. The mean is about 20 ppm and a range of 1-50 ppm occurs in agriculturally productive soils.[542] Much higher values may be encountered in soils derived from mineralized parent material.[80]

Because of the variety of conditions that influence the metal's availability, the total copper content of the soil is not an accurate indication of deficiencies or excesses of copper in soil-rooted plants. Copper is much more available to plants in soils with impeded drainage, because of the microbial activities associated with these conditions. In highly organic soils such as the peats or mucks, copper uptake by plants is usually low. It is especially limited in those soils with pH values of 6 and above. Appropriate weak extractant solutions such as a solution of ethylenediaminetetraacetic acid (EDTA)* may be used to obtain measures of available copper in soils. Paradoxically, copper deficiency may also occur in plants where the soil is acid if the solubilized copper salts are leached out. Deficiency also tends to occur in soils low in clay or high in organic matter. Copper toxicosis is uncommon in plants, but may be found in areas contaminated by mining or smelting activity.[69]

Special problems, such as the addition of copper salts to control alternate hosts for parasites or algal growth, may exist where ionic copper is present in the sea or in fresh water.[22,394] Surface and ground waters are potential sources of copper imbalances, particularly when added to soils through irrigation or flooding.[12,136,147,155,310,611,625] Nationally, the ecologic impact of water as a source of copper is difficult to evaluate because the necessary information is lacking. Above average levels of copper in water may be either natural or caused by man.[278,290,341,431,445,503,557] Most imbalances are localized near their point sources, whereas readily available information about the concentration of copper in water is for streams draining broad regions of about 10,000 square miles (26,000 km²).[290,551] Drinking water commonly is a minor source of copper, contributing less than 2% of the daily dietary intake.[218] For continuous use in irrigation, 1 ppm copper in water has been established as a maximum.[161]

Discussion of the significance of copper in soils and soil waters follows in Chapter 3, "Copper in Plants," and Chapter 4, "Copper in Animals." Such movement is of particular ecologic concern because of the varying sensitivities of different species of fish. Considerable attention has also been focused upon copper in animals and in animal wastes because of evaluations made of the movement of copper through the ecosystem from source to soil. Copper travels through water to freshwater and marine organisms, and to plants, animals, and humans.[29,90] Although

---

*Formulas for the major compounds mentioned are supplied in Appendix B.

some attention has been directed to human tissue concentrations,[496,555] most research has been directed toward the effect of sewage and animal and industrial waste disposal. Animal and industrial wastes (including sewage solids) commonly yield high concentrations of copper and other trace elements. The current emphasis on recycling these wastes may unintentionally supply excessive amounts of copper and these other elements to the soil.[121,310,416,625] Such recycling could indirectly affect consumers if the yield of crops were reduced or if copper were increased in feed products.

3

Copper in Plants

Copper has been known to be essential for certain fungi since 1927,[43] and since 1931 for the normal growth and development of green plants.[500, 603] Its nutritional essentiality has since been demonstrated for scores of plants, including all of the major agricultural crops.[152] Although the quantitative requirements of plants for copper are very low—only molybdenum is required in smaller amounts—there are numerous instances of naturally occurring copper deficiency. Copper toxicosis in plants is almost never observed under natural conditions, but may occur on mine spoils or where copper-rich manures or fungicides have been used excessively.

COPPER PROTEINS

The Blue Oxidases

It is well established that the copper-containing blue oxidases—ascorbate oxidase, the tree and fungal laccases, and the mammalian ceruloplasmin—contain at least two kinds of cupric copper.[331] Type I is associated with a sharp electron paramagnetic resonance (EPR) spectrum and an intense absorption in the visible spectrum around 600 nm, characteristics for which asymmetric coordination to the protein or charge transfer may be responsible. Type II copper is associated with a broader EPR spectrum and has no detectable absorption in the visible region. Anion binding that inhibits enzymatic activity will not affect the visible or EPR spectrum of Type I copper, but it will change the EPR spectrum of Type II copper.

Of the remaining copper atoms in the blue oxidases, some are possibly cuprous and a third type of multivalent copper may exist. Anaerobic redox titrations monitored by absorption at 330 nm (present in all of these proteins, in addition to the absorption in the region of 600 nm) and by EPR have yielded evidence that these proteins will accept one electron equivalent per copper atom. The amount of EPR-detectable copper is not sufficient to account for the number of electrons accepted. At least two alternative interpretations of this observation exist: a Cu^{+2} Cu^{+2} couple in which the paired electrons preclude detection by EPR spectroscopy or magnetic susceptibility; or, trivalent copper, i.e., Cu^{+3}, which Hamilton has suggested exists in galactose oxidase.[220]

Laccase

Laccase has been known for over 90 years. It was first isolated in 1883 from the milky secretion *urushi,* or latex, of the Japanese lac tree, *Rhus vernicifera.*[632] Laccases have since been found in many plants and fungi.[176] The animal protein ceruloplasmin[63] shares with laccases the capacity to catalyze the oxidation of a number of diamines[277,419] and diphenols,[535] including catechol, hydroquinone, and *p*-phenylenediamine. Diphenols and monophenols, present in coniferyl alcohol, may constitute the natural substrates of laccase.[180]

In addition to *Rhus* laccase, laccase from *Polyporus versicolor* has been widely studied. Both *Rhus*[382,449] and *Polyporus* laccases[62,149,333,376] contain four atoms of copper per molecule of enzyme. Both enzymes are characteristically blue in color, but will suffer loss of color, activity, and copper content by dialysis against cyanide. The molecular weight of *Rhus* laccase is approximately 120,000; that of *Polyporus* laccase is about 60,000.

For its forms of copper present and their roles in the catalytic activity of the enzyme, *Polyporus* laccase is perhaps the best understood copper enzyme. The copper system in *Polyporus* laccase is similar to that of other laccases and serves as the model system for all the blue oxidases.[327]

Nakamura[383] has proposed the following equations as representative of the laccase reaction:

$$2Cu^{+2} + \text{hydroquinone} \rightarrow 2Cu^+ + p\text{-quinone} + 2H^+$$

$$2Cu^+ + \tfrac{1}{2}O_2 + 2H^+ \rightarrow 2Cu^{+2} + H_2O.$$

In this "valence shuttle" hypothesis, copper in the oxidized enzyme is reduced first by one substrate (here hydroquinone), and the reduced enzymatic copper is then oxidized by the second substrate, molecular oxygen; each step requires a two-electron transfer. Studies with ceruloplasmin[63] have indicated a reaction sequence in which the substrate is oxidized in a one-electron step. This one-step reaction leads to a semiquinoid type of free radical demonstrable by EPR techniques. The same radical has been shown to be involved in oxidations with *Rhus* laccase[381] and ascorbate oxidase.[630]

Aside from pink cytochrome oxidase, the blue oxidases are the only oxidases in which water is the product of the enzyme reaction in oxygen reduction. Enzymes that catalyze the reduction of oxygen to hydrogen peroxide in a two-electron reduction are much more prevalent.[55,358,385] Therefore, the detailed mechanism for the four-electron transfer in the reduction of oxygen to water, and the way in which this reaction may be coupled to the one-electron oxidation of substrate, remains an unsolved problem. It has been discussed by Malkin and Malmstrom.[327]

On the basis of available data,[162,328,332,333] it is highly probable that the blue oxidases accomplish the four-electron reduction of oxygen to water in multielectron steps, most likely by double-electron transfers. Such a mechanism can only occur where cooperation exists between the different electron-accepting sites present in a single molecule of enzyme.

Ascorbate Oxidase

Ascorbate oxidase is another of the copper-containing blue oxidases. This enzyme has been found in many plants.[4,16,93,185,205,217,222,282,321,344,345,391,411,412,413,446,465,486,592,599] The richest sources are the yellow crookneck squash (*Cucurbita pepo condensa*) and the green zucchini squash (*Cucurbita pepo medullosa*), from which the most highly purified enzyme samples have been obtained.[123,308] The enzyme was first observed by Szent-Györgyi[544] in 1928, but it was not sufficiently purified to justify its classification as a copper protein until 1940.[317] Ascorbate oxidase catalyzes the oxidation of L-ascorbic acid to dehydroascorbic acid. It also catalyzes the reduction of cytochrome *c* in the presence of ascorbic acid and oxygen,[629] and is thought to play an important role in the electron transfer system of plants.[123–127,317,508]

Ascorbate oxidase has a molecular weight of 140,000[509] and a copper content corresponding to 8–10 atoms of copper per molecule of enzyme.[123, 308, 384, 532] There are two identical subunits of molecular weight of approximately 70,000, each of which is composed of two polypeptide chains of approximate molecular weights of 30,000 and 40,000.[532] Copper can be removed easily from ascorbate oxidase by dialysis against cyanide[421] to form the inactive, unstable,[294] copper-free apoenzyme. Copper and activity may be restored to the apoenzyme by forming the holoenzyme[89] with a molecular weight of about 285,000.[294]

Plastocyanin

Plastocyanin, a ubiquitous protein that contains two atoms of copper per molecule, plays a significant role in the electron transport system of plants undergoing photosynthesis.[36, 237, 271–276] Its concentration in chloroplasts is comparable to that of cytochrome f.[273] It is involved in the light-driven reduction of oxidized nicotinamide adenine dinucleotide phosphate.[270]

Vernon et al.[589] have shown that plastocyanin increases the rate of the dark-reduction of P700 (a small cytochrome-containing particle extracted from chloroplasts by Triton X-100) in the presence of ascorbic acid and indophenol dye.

Hind[236] has demonstrated that plastocyanin increases the rate of oxidation of endogenous cytochrome f through photosystem-I. There is an interaction between P700 and plastocyanin in both instances.[236, 589] A plastocyanin-lacking mutant of *Chlamydomonas* requires plastocyanin *in vitro* to mediate the electron flow between cytochrome f and photosystem-I.[206, 207]

Tyrosinase

Tyrosinase, or polyphenol oxidase, catalyzes the formation of melanin pigments in many plants and animals through the oxidation of tyrosine. It also catalyzes the oxidation of several monophenolic substrates and of o-dihydric phenols or catechols.[64] It was the first enzyme in which copper was shown to be an essential part of the active molecule.[296]

Tyrosinase displays two activities: catecholase activity, which involves the dehydrogenation of catechols, and cresolase activity, which involves the hydroxylation of monophenols. Whether these two activities both involve a single molecule with different active sites, or two different molecular species, or whether cresolase activity results from the hydroxylation of monophenols by the orthoquinoid product of the catecholase activity is a subject of considerable controversy.[64,280,343]

Copper content and activity may be restored to the apoenzyme of tyrosinases by treatment with Cu^{+2}.[44,280,296] The cupric ion added during reconstitution is reduced to the cuprous state; any excess copper remains cupric.[280]

Potato and mushroom tyrosinases (molecular weight 120,000) have been found to contain four atoms of copper.[44,296,329] All of the atoms are thought to be cuprous.[279,281]

Amine Oxidases

These enzymes are copper proteins that catalyze the oxidation of amines, as indicated by the following equation:

$$RCH_2NH_2 + O_2 + H_2O \rightarrow R-\overset{\displaystyle O}{\overset{\|}{C}}-H + NH_3 + H_2O_2.$$

They generally are pink and are found in a variety of plants, as well as in animals and bacteria. Most of the copper in these enzymes is in the cupric state. A number of reviews have been published about amine oxidases.[37,385,634-636]

Stellacyanin

Stellacyanin (*Rhus* blue protein), like laccase, is found in the Japanese lac tree, *Rhus vernicifera*. This copper protein has a molecular weight of 16,800 and contains one atom of copper per molecule bound to a highly asymmetrical site.[400,401] Stellacyanin contains 20% carbohydrate and 20% hexosamine, in addition to its 108 amino acid residues.[420]

The biologic function of stellacyanin is not known; it may function as an intermediary electron carrier.[420] It does not possess oxidase activity.

COPPER DEFICIENCY

Copper occurs in plant material at concentrations of 1–50 μg/g dry weight of tissue; average concentrations are listed in Table 3-1. Except for molybdenum, it is the least abundant substance among the known essential plant nutrients.[28,468,529] Concentrations below 5 μg/g are a likely sign of deficiency.[260,450] The extremely sparse occurrence of copper corresponds to low quantitative requirements for the metal, but deficiencies have been reported.[297,379,450]

TABLE 3-1   Copper Concentrations in Representative Agricultural and Horticultural Plant Species[a]

Plant or Plant Part Analyzed	Number of Analyses	Copper content, ppm (dry wt)		
		Maximum	Minimum	Mean
Alfalfa, aboveground portion cut for hay	8	15	4	9
Barley, grain	12	41	6	16
Beans, field; seed	12	16	7	11
Beets, root	15	27	6	10
Cabbage, edible portion	26	28	4	14
Carrots, root	15	18	7	11
Clover, red, aboveground portion	41	20	6	10
Corn, grain	6	17	4	8
Corn, stover	16	9	2	5
Kale, edible portion	6	56	24	36
Lettuce, edible portion	45	33	3	19
Oats, grain	29	51	4	11
Oats, straw	26	54	3	11
Onions, bulb	11	24	5	12
Orange, fruit	3	22	3	10
Peas, green, edible portion	9	15	6	9
Potatoes, tuber	143	24	2	8
Soybean, aboveground portion cut for hay	32	12	4	9
Spinach, edible portion	34	24	3	9
Tomato, fruit	51	34	8	14
Wheat, grain	108	24	4	9
Wheat, straw	24	5	1	3

[a] Data from Beeson.[28]

The limiting concentration of copper in the soil water is of the order of 0.5 mg/kg, or about 6 mg/kg of water and solids.[283] High concentrations of humus or heavy additions of lime result in complexing of copper with high-stability constants and cause the limiting concentrations of copper to be varied upward to about 30 mg/kg. High concentrations of phosphate, manganese, or zinc can also result in potential copper deficiency; the metals apparently compete directly with copper for transport sites on the plant roots. Soils containing a great deal of organic matter that are brought into agricultural production for the first time frequently trigger copper deficiency—hence the term "reclamation disease."* Crops in deficient regions can be successfully treated with cupric sulfate, copper EDTA, copper lignin sulfonates, or copper flavonoids, which usually are incorporated into the soil by spraying. However, the leaves may be sprayed[379] or the seed may be treated instead. Actual dosages are about 3 kg copper/ha.[379]

The amount of copper in plants also depends upon microbial activity of the soil, pH, oxidation-reduction potential, moisture, rainfall, and, of course, the species of the plant.[10, 11, 28, 163, 543, 560, 572]

Grasses, averaging 5 ppm, tend to be lower in copper than are legumes, which average 15 ppm. Grains (seeds) tend to be higher than leaves or stems: Animals on diets consisting principally of whole grains rarely develop copper deficiencies. Regulatory mechanisms appear to limit the concentration of copper found in plant tissues to about 20 ppm, although higher levels occur under specialized conditions in some plant species[66] and in seeds.[440]

The signs of copper deficiency in fruit trees are known as "exanthema" or "dieback" and include death of the apical bud, rosetting and multiple bud formation, and chlorosis (yellowing) of the leaf margins. In cereals, the younger leaves wither, show marginal chlorosis, and often fail to unroll. The flower heads are dwarfed and the tips are chlorotic and underdeveloped, yet the lower leaves remain green and bushy. Grain formation is disproportionately affected.[297, 603]

The extreme susceptibility of bud development and the failure of an otherwise fairly normal vegetative plant to set seed are symptomatic of weak translocation of copper from various parts of the plant to the bud. Similar patterns are observed with several plant nutrients, such as boron. Others, notably nitrogen, are preferentially transported from older to younger tissues. The redistribution of copper from leaves to stem, bulbs, and other storage tissues has been measured.[553]

Although there are no outstanding problems at the whole-organism level concerning the requirements of higher plants for copper, our understanding of the physiologic and molecular bases for these requirements is unsatisfactory. Nonetheless, numerous examples attest to the beneficial effects of copper on a number of crops.[556]

When fertilizer was supplemented with 0.3–0.8 g copper/m^3, eggplants took longer to bloom, but their yield increased 23%.[7] The application of cupric sulfate and ammonium molybdate (10 and 5 mg/kg soil, respectively) to spring wheat increased its yield by 12–27%.[404] Treating it with cupric sulfate before sowing increased the yield of grain, the fullness of the wheat seedlings, and the percentage of wheat plants

---

*A copper-deficiency disease to which many crops, especially cereals, are susceptible, usually occurring on newly reclaimed peat land and characterized by chlorotic leaf tips and failure to set seed.

surviving to harvest.[533] Cotton seed treated with 0.01% cupric sulfate or soil application of copper (10–20 mg/kg soil) accelerated the maturation and increased the yield of cotton[633] and cotton oil.[444] When apple trees were sprayed with aqueous copper solutions, the productive photosynthesis of the leaves increased markedly, growth was stimulated, and carbohydrates accumulated.[291,488] Young corn plants increased their rates of photosynthetic oxygen production when copper was added.[179] Treating tobacco seeds with 0.01% cupric sulfate increased the activity of iron and copper proteins (catalase, ascorbate oxidase, polyphenol oxidase, and peroxidase), and the yield of tobacco.[444] The yields of legumes[422] and dorset marlgrass[235] similarly increased when their seeds were treated with 0.01% cupric sulfate.

COPPER TOXICITY

When swine and poultry are raised on diets rich in copper (250 ppm),* the high level of copper in their manure may significantly increase copper in soil.[120,638] The level required to affect plant growth adversely will depend upon the content of the clay and organic matter. Sandy soils may reach such a level in 5 years,[451] whereas silt, loam, or peat soils may not show adverse effects for 50–100 years. Widespread and repeated use of copper compounds as fungicides also has resulted in soil accumulations in certain areas.[133] The application of sewage sludge also affects the copper content of soils and of plants growing in them. Leeper[309] has reviewed the effects of copper upon the microorganisms active in sewage digestion, as well as the effects of sewage sludge containing up to 8,000 ppm copper upon soils and the crops grown in them. The copper did depress microorganism activity, but recovery was rapid.

* See Chapter 4.

4

Copper in Animals

Copper has varied and numerous biologic effects in animals as an essential element as well as a toxicant. Much of the knowledge of the metabolism, physiology, and biochemistry of copper in man presented in Chapter 5 applies to other animals (especially mammals) as well. This chapter will treat principally those aspects of the animal biology of copper that differ from the human.

Animal tissues show a wide range of copper concentration. To some extent, this range reflects the level of copper in the diet, especially when food with excessively high or low copper content succeeds in defeating the animal's homeostatic processes. Whole body copper contents of most animals on average diets range around 2 ppm in the fat-free tissues. Highest concentrations are found in the liver and brain, with lesser amounts in the heart, spleen, kidneys, and blood.[87,88,111,319,496,504,571,614]

Brain tissue appears to be the only tissue in which copper concentration increases with age, approximately doubling from birth to maturity.[480] An exceptionally high concentration is found in the pigmented portions of the eye, particularly the iris and choroid, where amounts up to 100 ppm dry weight can occur.[49,50] The amount of copper in serum can range from 5 to 130 μg/100 ml.

The occurrence of disorders related to either a deficiency or an excess of copper in the United States is considered in detail in this chapter and the next.[295]

COPPER DEFICIENCY

The actions of copper at the cellular level generally involve copper proteins, many of which are enzymes with oxidative functions. Probably no metal ion is more versatile than copper as a requirement of specific enzymatic reactions. Tyrosinase, laccase, ascorbic acid oxidase, uricase, monoamine oxidase, dopamine-β-hydroxylase, and cytochrome oxidase have all been identified as copper enzymes;[571] indeed, diminished activity of cytochrome oxidase is a sensitive indicator of copper deficiency.[264]

Gallagher and Reeve[187] have suggested that an uncomplicated copper deficiency in the rat first causes the loss of cytochrome oxidase activity. This loss leads to depressed hepatic mitochondrial synthesis of phospholipids because it interferes with the provision of sufficient endogenous adenosine triphosphate (ATP) to maintain an optimal rate of synthesis. Although copper is involved in many other biochemical functions, depressed phospholipid synthesis is probably the primary result of insufficient cytochrome oxidase.

Severe anemia is a prominent manifestation of copper deficiency in swine and other animals.[85,301,571] Copper deficiency is first manifested by a slow depletion of body copper stores, including blood plasma. The type of anemia associated with copper deficiency is identical to that caused by iron deficiency.[85,215,301,307] Moreover, animals fed a diet deficient in copper yet given adequate amounts of iron orally fail to absorb iron[91,92,215,216,306,307] and are iron-deficient. Such animals even fail to respond to parenteral iron.[85,306,307] Their mucosal epithelial cells, hepatic parenchymal cells, and reticuloendothelial cells are able to take up iron normally, but they are unable to release iron to the plasma at the normal rate.[306,307,443,456]

In pigs, ceruloplasmin appears to be essential for the movement of iron from cells to plasma,[405] and lack of this copper protein accounts for more than defects in iron metabolism. Reticulocytes from copper-deficient animals can neither take up iron from transferrin normally nor synthesize heme from Fe (III) and protoporphyrin at the normal rate.[616] Mitochondria from copper-deficient animals lack cytochrome oxidase, which apparently is required to reduce Fe (III) to Fe (II) to provide a pool of Fe (II) as substrate for heme synthesis. Thus, there are multiple defects in iron metabolism in copper-deficient animals, and the copper enzymes ceruloplasmin and cytochrome oxidase are intimately associated with the movement of iron.

A variety of disorders in animals and man have been associated with copper deficiency, but the concentration of copper relative to molybdenum, zinc, iron, and sulfate is essential in defining the clinical significance of the deficiency.[342] Indeed, the ratio of copper to these dietary components appears to be almost as important as the actual level of copper in the diet. The pathogenesis of and susceptibility to copper deficiency are different in ruminant and nonruminant animals, because in ruminants the interactions of molybdenum and sulfate ion with copper are of primary importance, whereas in nonruminants the interactions of iron and zinc with copper predominate.[499]

Ruminant Animals

Disorders associated with a relative copper deficiency in various ruminants include anemia, depressed growth, bone disorders, depigmentation of hair and wool (achromotrichia), abnormal wool growth, neonatal ataxia, impaired reproductive performance (fetal death and resorption in rats; depressed estrus in cattle), heart failure, cardiovascular defects, and gastrointestinal disturbances.[571] Many factors influence the severity of these dysfunctions, especially species, and even breed or strain characteristics, age, dietary interrelationships, environment, and sex.

Bone abnormalities associated with copper deficiency have been reported in rabbits, mice, chicks, dogs, pigs, foals, sheep, and cattle.[571] In ruminants, osteoporosis and spontaneous bone fractures are usually associated with excess dietary molybdenum and a relative copper deficiency, but Suttle et al.[538] have presented evidence of the development of osteoporosis in the offspring of ewes fed diets totally deficient in copper.

Sheep suffering from simple copper deficiency and/or excess molybdenum also develop depigmentation of dark wool as well as loss of crimp and quality of their fine wool.[137,571] In Australia, a syndrome called enzootic ataxia and in the United Kingdom a condition termed swayback are probably caused by copper deficiency. Ewes with enzootic ataxia become anemic. Their wool is stringy and their lambs develop neurologic problems. Swayback lambs—particularly those less than a month old— are severely uncoordinated, ataxic, and usually blind, but the ewes' wool is normal. Death comes from starvation, exposure, or pneumonia.[264,571] Cordy[104] has reported that enzootic ataxia also occurs in the United States.

Pathologic lesions associated with enzootic ataxia and swayback in lambs include myelinolysis of the white matter of the cerebrum and degeneration of the motor tracts of the spinal cord. The destruction of the white matter may range from microscopic foci to massive subcortical destruction. Neuronal degeneration as well as demyelination often results.[104,264]

The first evidence of cardiovascular disorders caused by copper deficiency emerged from studies by Bennets and co-workers[30-32] of a disorder in cattle known as falling disease. Sudden deaths characteristic of the disease were attributed to heart failure, usually after exercise or excitement. A similar condition, preventable by copper supplements, also has occurred in pigs and chickens,[396,487] but it has not been reported in sheep or horses.

In the cardiovascular disorder, there is derangement of the connective tissue in major blood vessels, and spontaneous ruptures result. The tensile strength of the aorta is markedly reduced and the myocardium becomes friable. The primary biochemical lesion has been described by Hill *et al.*[234] as reduced activity of lysyl oxidase, a copper-containing enzyme, in the aorta. This reduction in enzymatic activity diminishes the capacity for deaminating lysine in collagen. Consequently, fewer lysine ϵ-amino groups are converted to an aldehyde function, cross-linkage is diminished, and tensile strength is reduced.

Apparently cattle are more susceptible than sheep to the combination of excess molybdenum and deficient copper in their diet.[58,175,571] When the ratio of copper to molybdenum in feed drops below 2:1, clinical manifestations attributed to molybdenum poisoning (but just as logically to copper deficiency), can be expected in cattle unless the copper concentration in the feed exceeds 13–16 ppm. This syndrome is manifested by emaciation, liquid diarrhea full of gas bubbles, swollen genitalia, anemia, and achromotrichia. Prolonged purgation may inhibit weight gain and cause death. About 80% of the cattle fed this diet develop molybdenosis; if cases are not treated, the fatality rate may be equally high.[58] Osteoporosis and bone fractures have been reported in prolonged cases of molybdenosis.[571]

Feeding cattle forages and grains grown on soils naturally high in molybdenum and/or low in copper also brings on this condition.[552] In the United States, such soils have been found in California, Oregon, Nevada, and Florida.[58,175,571] Cattle grazing in pastures on muck or shale soils in England, Ireland, New Zealand, and the Netherlands also have suffered severe molybdenosis.[571]

Miltimore and Mason[363] have made an extensive report of molybdenum and copper concentrations and copper:molybdenum ratios in ruminant feeds. The overall mean copper:molybdenum ratio of all feeds—legume hay, grass hay, sedge hay, oat forage, corn silage, and grains—was 5.7:1 in British Columbia. The copper:molybdenum ratio in sedge hays was 2.1:1, near the critical ratio of 2:1 that will lead to copper deficiency in cattle. The mean ratio of other hays was 4.4:1, and the ratio for other feeds was over 5:1. Molybdenum levels were generally low—often less than 1 ppm. Most copper levels were below 10 ppm.

When the copper levels of feed or forages are normal (ranging from 8 to 11 ppm), cattle generally are resistant to molybdenum poisoning from feed containing molybdenum levels as high as 5-6 ppm. Sheep can resist levels up to 10-12 ppm. But when the dietary copper level falls much below 8 ppm, even 1-2 ppm molybdenum may be toxic to cattle. Increasing the copper level in the diet to 13-16 ppm will protect cattle against 150 ppm molybdenum[137,571] if adequate sulfate ion is present, although the critical ratio of a normal diet—one with 8-11 ppm copper and 1-2 ppm molybdenum—is 2:1. In certain areas of the United States, such as Florida and states west of the Rocky Mountains, it is not uncommon to find molybdenum-induced copper deficiencies in cattle and occasionally in sheep.[33,96,104] Copper deficiencies in plants and animals are unusual in most areas east of the Rocky Mountains. Because there is no national program for reporting cases of copper deficiency in animals, it is not possible to define the extent of this problem.

Nonruminant Animals

Copper deficiency in nonruminant animals will result in anemia, aortic rupture, bone deformation and reduced calcification, cerebral edema, cortical necrosis, achromotrichia, and fetal absorption.[571] The levels of ceruloplasmin and copper in serum and of cytochrome oxidase and copper in tissues all decrease in animals fed a copper-deficient diet.[144,251,409,443]

Nonruminants are more tolerant of excessive levels of molybdenum than ruminants. Swine appear to be the most tolerant of the nonruminants. Davis[121] reported that a diet containing 1,000 ppm molybdenum fed to swine for 3 mo had no ill effects on them. In pigs, the storage of copper in the liver does not appear to be influenced by the level of molybdenum in the diet.[229,287,288]

However, excess molybdenum in rats may cause symptoms similar to those produced by a copper deficiency, and, as in ruminants, the level of molybdenum required to produce toxicosis depends upon the amount of copper in the rat's diet. Neilands et al.[388] and Gray and Daniel[211] have demonstrated that growth rate and hemoglobin levels of rats on low copper diets can be reduced by feeding them 100 ppm molybdenum. When the diet contains adequate amounts of copper, 500-1,000 ppm molybdenum is required to cause such effects. In rats, copper levels in the blood and liver tend to increase when molybdenum is added to their diet.[98]

Supplemental L-ascorbic acid has been demonstrated to aggravate copper deficiency in chicks, swine, and rabbits.[193] Hunt et al.[251] found that 0.5% dietary ascorbic acid reduced hepatic copper levels and increased deaths caused by ruptured aortas. Voelker and Carlton[593] have reported that 2.5% ascorbic acid in the diet of swine intensified symptoms of copper deficiency.

Numerous reports[17,387] have indicated that swine fed diets high in copper (up to 250 ppm) during the 8 wk of early post-weaning period increased their daily weight gains. However, continuous feeding at these levels did not affect overall rate of gain or feed-gain ratios. Copper stores in the liver increased linearly with increasing dietary copper, but removal of the added dietary copper reduced hepatic copper content. Marked differences were discovered from location to location, suggesting that genetic and environmental factors may contribute to the differences observed in swine. The action of the high level of copper in promoting early rather than later growth in animals fed the metal resembles the action of broad-spectrum antibiotics. Perhaps the copper exerts antimicrobial action in the intestine.[74,75]

The difficulty in evaluating the effects of feeding swine high levels of copper was revealed by the North Central Regional Committee Report.[387] Feeding 125–250 ppm copper to swine for up to 8 wk after weaning caused an average increase in daily gain over no added copper, although wide variations did occur from state to state.

Such variations resemble those reported for the use of broad-spectrum antibiotics like Aureomycin. When pathogen levels were significant in the environment, addition of the antibiotic significantly improved the daily rate of gain. When the environment was relatively free of pathogens, response to antibiotics was minimal.[110]

Although tests by the Food and Drug Administration (FDA) failed to demonstrate its effectiveness, widespread use of poultry rations supplemented with 250 ppm copper sulfate is generally believed to prevent or cure crop mycosis (moniliasis caused by *Candida albicans*).[573] When used in conjunction with antibiotics, the net effect of copper sulfate is to reduce the growth rate of poultry. [47,493]

As with swine, results vary when high levels of copper are fed to poultry. A level of 240 ppm copper given turkey poults encouraged growth in Iowa,[6] 60 ppm copper in the diet of chicks in Ontario was without effect,[493] and 125 ppm in the diet of young turkeys raised in Florida depressed their growth.[226]

COPPER TOXICITY

Copper toxicity and interactions of copper with other trace elements present complex and significant consequences for animal husbandry in the United States. Dietary imbalances of copper and molybdenum may result from *ad libitum* consumption of either mineral mixtures or of conventional feeds that have been fortified with inappropriate mineral mixtures.[45,70,72,292,293,541]

Up to 15 ppm copper is generally recognized as safe (GRAS) in livestock feed by the FDA, whereas molybdenum is not considered a GRAS substance at any concentration. Therefore, copper is routinely and ubiquitously added to commercial trace element mixtures used in livestock feeds, whereas molybdenum is prohibited. Unfortunately, these regulations do not recognize species differences between cattle and sheep in their requirements for a balance between copper and molybdenum. Cattle can tolerate mineral mixtures and feeds with added copper and without molybdenum, even when their natural grain and forages contain adequate levels of copper. In contrast, sheep are susceptible to the toxic effects of added copper, especially when the natural forage contains adequate levels of copper and low levels of molybdenum. Since the cattle-feeding industry is of major economic importance, and the sheep-feeding industry is not, it has not been economically feasible for manufacturers of livestock mineral mixtures to provide special formulations for sheep with the proper balance between copper and molybdenum (6-10 parts copper/1 part molybdenum).

Copper toxicoses in sheep are not rare in the Midwest and Great Plains states. They extend northward well into Canada. Because the levels of copper found in plants vary greatly and depend upon many factors,* no general geographic distribution of copper and molybdenum levels in plants has been mapped. However, it appears that grains and forages grown in the upper Midwest and Great Plains states contain sufficient copper and are low enough in molybdenum content to make the addition of the GRAS 15 ppm copper to the total diet of sheep responsible for excessive accumulation of copper in the liver. In these areas, 1-5% of sheep consuming such feed develop hemolytic crises. Sheep may develop copper toxicosis on a diet containing a normal concentration of copper (8-10 ppm) if the molybdenum levels are below 0.5 ppm.[71,239] When a vitamin–mineral preparation containing copper but not molybdenum is added to a ration, the copper concentration of the ration may be elevated to 25-30 ppm or more. Since the natural molybdenum concentration in most feedstuffs is usually below 2 ppm, the copper:molybdenum ratio in the resulting diet is greater than 10:1. Over 20 episodes of chronic copper toxicosis in sheep were found in Iowa from June 1968 to June 1970,[70] especially in feeder lambs, show lambs, and ram lambs being

* See Chapters 2 and 3.

tested for weight gain and feed efficiency (unpublished data, W. B. Buck).

These problems could be solved economically[390] by computerized on-line feed-forward control of feed mills. For example, on-line analysis of incoming ingredients for their copper, molybdenum, sulfate, iron, and zinc contents would produce data from which to calculate the trace elements necessary to provide the proper levels and balance in feeds for each species.

Adding excess copper to sheep, swine, and poultry feeds may create a hazard for the consuming public because the metal accumulates in the animal liver.* Sheep fed a diet with copper and molybdenum imbalances have accumulated an average of 1,600 ppm copper in the liver tissue on a wet-weight basis.[15] In some instances, accumulations have run as high as 3,000 ppm.[3,564] Pigs fed rations containing 250 ppm copper had a mean hepatic concentration of 220 ppm on a dry-weight basis, compared to a mean of 24 ppm in animals not receiving the added copper. Even higher levels of copper may accumulate in livers of swine if their dietary levels of iron and zinc are inadequate. Using sheep, swine, and poultry livers from animals fed such diets for human consumption could be deleterious, especially in the preparation of baby food. Baby foods made from liver containing 550 ppm copper (wet wt) would contain 15 mg copper/1-oz (28 g) serving. This level of copper is 5–10 times the daily dietary requirement for girls aged 6–10 yr.[151]

Other causes of copper toxicosis in ruminants include:

- consumption of plants contaminated by copper-containing pesticides used to spray orchards, such as a Bordeaux mixture with 1–3% copper sulfate;
- use of copper sulfate to control helminthiasis and infectious pododermatitis in cattle and sheep;
- contamination of soils and vegetation in the vicinity of mining and refining operations;
- use of calcium–copper EDTA as an injectable source of copper in countries where sheep frequently are subject to copper deficiency problems;[254,255,258] and
- confining sheep[3,52,558] with no access to green forage containing sufficient molybdenum to prevent excessive accumulation of copper in the liver.[560]

In western Australia, sheep grazing on pastures containing various

*See Chapter 5.

species of *Lupinus* develop hepatic toxicosis from lupine alkaloids more readily in the presence of excess copper.[188,189] Susceptibility to copper poisoning in sheep may be enhanced by the forage. Thus, in Australia and New Zealand, plants of the *Heliotropium, Echium,* and *Senecio* genera contain pyrrolizidine alkaloids that cause hepatic necrosis; animals grazing on these plants will be unable to metabolize and excrete normal dietary levels of copper.[27,73,461,571]

Toxic Interactions

Before discussing the pathologic physiology of copper toxicosis, it is essential to review the mechanism of interaction of copper with molybdenum, sulfate, iron, and zinc. Evidence exists that copper and molybdenum form an *in vivo* complex with a molar ratio of 4:3,[145,146,249] which may not prevent intestinal absorption of copper, but which does inhibit copper accumulation in the ruminant liver.

Copper and molybdenum, especially in ruminants, appear to interact with inorganic sulfate in the diet,[112,138,139,239,337,559] affecting biliary and urinary excretion of copper and molybdenum. Dick[139] reported that increased urinary excretion of molybdenum occurs with increased content of inorganic sulfate in the diet, and Marcilese *et al.*[337,338] found that increased dietary levels of molybdenum and sulfate result in increased urinary and biliary excretion of copper. Huisingh and Matrone[249] found that molybdate inhibited the reduction of sulfate to sulfite, and that copper reduced this inhibition greatly. Molybdate inhibition of sulfate reduction increased as the concentration of sulfate decreased.

Copper added to sheep diets in the sulfate form may be less toxic than copper added as the acetate, oxide, carbonate, gluconate, iodide, chloride, orthophosphate, or pyrophosphate.[70,71,561]

The first clinical evidence of the relation between copper and molybdenum metabolism was obtained when it was learned that teart, the drastic scouring disease of cattle—thought to be a manifestation of chronic molybdenum poisoning—could be controlled by treating the cattle with large amounts of copper.[168,571] Then Dick and Bull[140] reported that molybdenum was an effective treatment for copper poisoning in sheep. Subsequent investigations have characterized these interactions in ruminants.[3,112,137-139,145,146,204,248,292,293,337,428,459,558,559,571]

The effects of interactions of copper, molybdenum, and sulfate are much less marked in the nonruminant.[105] Gipp *et al.*[194] and Hays and Kline[229] were unable to demonstrate any effect of molybdenum and sulfate on the storage of copper in the liver by pigs fed varying levels of copper. Dale[114] observed similar results, although he found that ceruloplasmin levels were depressed when sulfate was added to swine diets containing about 10 ppm copper.

Zinc and iron affect copper metabolism in nonruminant animals, especially swine. Both elements protect swine from the adverse effects of high levels (250–750 ppm) of dietary copper,[224,454,539,540] and deficiency of zinc and iron tends to intensify copper toxicosis in swine.[540] In rats, copper was a prophylactic against the anemia[499] and reduced activities of catalase and cytochrome oxidase in the liver associated with zinc toxicosis.[210,586]

PATHOLOGIC PHYSIOLOGY OF COPPER TOXICOSIS

Ruminant Animals

Copper chloride is 2–4 times more toxic than copper sulfate and sheep are poisoned by 20–100 mg/kg single dose. Signs of acute poisoning by large oral doses of a copper formulation are vomiting, excessive salivation, abdominal pain, and diarrhea (fluid, greenish-tinged feces). Collapse and death follow within 24–48 h.

When sheep consume small but excessive amounts of copper over a period of weeks to months, particularly when the copper:molybdenum ratio is greater than 10:1, no toxic signs will be manifested until a critical level of copper—3–15 times the normal level, or about 150 ppm (wet wt)—is reached in the liver. Suddenly, the animal becomes weak, trembles, and loses its appetite. It usually develops hemoglobinemia, hemoglobinuria, and icterus. Occasionally, an animal will only show pale mucous membranes, and not icterus or hemoglobinuria.[537] Although morbidity is usually less than 5%, the mortality of the affected animals is usually over 75%.

The hepatocytes may exhibit cytoplasmic vacuolation and necrosis. All lobules may contain clusters of dead cells with fragmented nuclei and acidophilic cytoplasm. Fibrosis begins early and is distributed portally.[264] The kidney tubules are clogged with hemoglobin; accompanying degeneration and necrosis of the tubular and glomerular cells occurs. The spleen is crowded with fragmented erythrocytes, and status spongiosus in the white matter of the central nervous system has been reported.[257]

Morphologic and histochemical changes occur in sheep when copper accumulates in their livers.[257] In biopsies taken 6 mo before the hemolytic crisis, swelling and necrosis of isolated hepatic parenchymal cells have been noted, together with swollen Kupffer cells rich in acid phosphatase and containing p-aminosalicylic acid (PAS)-positive, diastase-resistant material and copper. Various increases in liver-related serum enzyme activities have been recorded 6–8 wk before the hemolytic crisis. These enzymes include serum glutamic oxaloacetic transaminase, lactic dehydrogenase, sorbitol dehydrogenase, arginase, and glutamic dehydrogenase.[256,257,322,458,562,564,584] The increased serum activities of these enzymes often subside to nearly normal levels 1–2 wk before the hemolytic crisis, but very high levels of activity occur shortly before or during the crisis. It is important to note that these elevations are not correlated with increases of copper levels in the blood, which only occur shortly before and during the hemolytic crisis and therefore are of no diagnostic value before the animals fall sick.[350]

During the hemolytic crisis, the activities of hydrolytic adenosine-triphosphatase, nonspecific esterases, and succinic dehydrogenase are reduced.[256,257,425]

Todd and Thompson,[562,563] working with sheep, reported a marked reduction of glutathione concentration and an accumulation of methemoglobin in the blood in the hemolytic crisis of copper toxicosis resulting from a sudden release of copper from the liver. Death may result from blockage of the kidneys by hemoglobin and subsequent kidney failure.

Nonruminant Animals

Toxic levels for ruminants (20–50 ppm) are well tolerated by nonruminants. Dietary levels in excess of 250 ppm are required to produce toxicosis in swine and rats.[51,539,540,602] Copper toxicosis in nonruminants may not cause rapid destruction of red blood cells—the hemolytic crisis—although jaundice has been observed in pigs fed toxic levels (250–500 ppm) of copper.[571,602] Milne and Weswig[362] have shown that sheep accumulate copper in the liver in proportion to the dietary intake, whereas rats maintain normal hepatic copper levels until a diet extremely high in copper (1,000 ppm) is reached.

Studies with rats and mice injected with copper compounds have shown that copper accumulates in liver lysosomes.[20,197,304] Some researchers have postulated that acid hydrolases capable of producing cellular injury are thereby released, thus causing hepatic damage.[313,314,588] Conversely, it has been demonstrated that high concentrations of copper in the toad, *Bufo marinus,* are localized to liver lysosomes and are made innocuous because of this localization.[199]

Copper levels of 125–250 ppm in the diet of swine increase the unsaturation of depot fat so that it turns soft.[129,150,547]

Poultry resist copper toxicosis better than most mammals.[357]

Smith[498] fed copper sulfate to day-old chicks for 25 days at zero, 100, 200, or 350 ppm copper concentrations in a basal ration containing 10 ppm copper. Chicks on the 100 ppm copper diet increased slightly in daily gains, whereas those on the 350 ppm diet showed a slight but statistically significant reduced weight gain. Goldberg *et al.*[196] gave copper acetate in capsules to adult chickens (weighing 1.8 ± 0.25 kg) at a rate of 50 mg copper per chicken/day for 1 wk, 75 mg/day for a second week and 100 mg/day until anemia or toxicosis appeared or death occurred. After 2-6 wk of being dosed with copper, the birds became weak, anorectic, and lethargic. Eight out of 23 developed anemia concomitant with toxicosis, perhaps because erythrocytes were destroyed in the liver by exposure to copper. Turkey poults have been reported to tolerate up to 676 ppm dietary copper sulfate for 21 days, but growth was depressed when fed 910 or 1,000 ppm copper, and signs of toxicosis appeared at 1,620 ppm.[597] Wiederanders[615] tried to produce copper toxicosis in turkeys by injecting copper sulfate subcutaneously. He injected 0.5 mg per bird for 84 days, and elevated the dose to 5.0 mg per bird for an additional 17 days, yet copper toxicosis was not produced. He concluded that turkeys and perhaps other fowl have metabolic and excretory pathways for copper different from those of mammals and pointed out that ceruloplasmin did not increase in the turkeys injected with copper.

Extensive studies of acute and chronic copper toxicosis in chickens, pigeons, and ducks were conducted by Pullar.[442] He found the minimum lethal dose (MLD) of copper for these species to vary from 300-1,550 on a mg/kg body weight basis, depending on the form of copper fed. The maximum daily intake of copper carbonate tolerated by chickens was 60 mg/kg body weight; mallards tolerated 29 mg/kg body weight. The greater sensitivity of ducks to copper toxicosis may be related to their tendency to accumulate more dietary copper in their livers than do chicks. It was not possible to produce poisoning in chickens given copper sulfate in drinking water at 250 ppm (1:4,000 dilution of copper sulfate in drinking water), and no obvious signs of copper poisoning were observed in mallards consuming 250 ppm copper sulfate in their feed.

Numerous studies in rats and mice have been conducted in an effort to learn more about hepatolenticular degeneration (Wilson's disease) in humans.[20,303,304,313,314,350,588,594,623] Prolonged daily intraperitoneal injections of as little as 0.3 mg copper/kg body weight will elevate hepatic copper levels.[303,304] There is indication that an increase in hepatic copper occurs without saturating the rat's excretory capacity. Copper levels in the kidney also increase with copper exposure, but this seems to be unrelated to liver storage.[303,304] Both hepatic and renal necrosis observed in rats and mice are linked to increased copper levels.[594,623] However, there is no apparent deposition of copper in the brain, skeletal and cardiac muscle, or skin, and only transient elevations of copper are found in bone following copper exposure.[303,304]

Hardy *et al.* have described a form of hepatic cirrhosis in Bedlington terriers that has striking similarities to Wilson's disease.[225] The disorder appears to be hereditary and autosomal, and the histologic and functional abnormalities of the liver are extraordinarily like those seen in man. The concentrations of hepatic copper found in these terriers exceed 10 μg/g (10,000 ppm) dry liver, to be compared to 0.25–3.0 μg/g in patients with Wilson's disease and normal levels of less than 0.1 μg/g in humans and dogs. Although this disorder is fatal to the dogs, Kayser-Fleischer rings and neurologic dysfunction have not been observed so far.

Aquatic Organisms

Copper is poisonous to many aquatic organisms. It may reach toxic levels from mining or industrial operations, influxes of copper-containing fertilizers, or use of copper salts to control aquatic vegetation or mollusks. Desalinization plants may cause local excessive concentrations of copper in the ambient salt water because their effluent is hot, hypersaline, and of low pH—conditions that will dissolve the metal in the copper pipes or vessels through which the waste flows. Toxic concentrations are also functions of the species, the age of the individual organism, the concentrations of mineral and organic material, the temperature of the water, and whether or not the copper is ionic.

In fresh water, acute toxicosis in fish is unusual if the concentration is below 0.025 ppm. (The accepted standard for drinking water is 1.0 ppm.)[147] In soft fresh water, however, 0.01–0.02 ppm has been found to be toxic.[261,262,427,437]

As exposure time is lengthened, the minimal toxic concentration diminishes. The 48-h LC_{50} (lethal concentration for 50% of the animals) in rainbow trout (*Salmo gairdnerii*) has been found to be 0.67–0.84 ppm.[67] The 96-h TL_m (medium threshold limit, killing 50% of test animals in 96 h) of copper for bluegills (*Lepomis machrochirus*) is reported as 0.24 ppm,[397] although levels over 0.01 ppm alter oxygen consumption. The 10-day LC_{50} of copper for brook trout (*Salvelinus fontinalis*) is about 0.05 ppm.[502] Chinook salmon eggs can withstand 0.08 ppm,[230] but the fry exhibit acute toxicosis at 0.04 ppm, and even 0.02 ppm copper inhibits their growth and increases mortality.

Relatively few data are available on more chronic exposure of fish to copper. Mount[377] found that fathead minnows (*Pimiphales promelas*) exposed to copper for 11 mo did not show impaired growth or reproduction at 3-7% of the 96-h TL_m of 0.43 ppm. That is, they were unaffected by 0.02 ppm. Minnows are unaffected by 3 times this concentration of copper if the water has an EDTA hardness of 30 mg/l as calcium carbonate.[378] In surprising contrast, Arthur and Leonard[12] found that 8 to 14.8 ppm copper had no effect on fish after 6 wk in soft water, but this tolerance might have been a phenomenon of species difference.

Copper pollution of waters has appreciable effect on marine invertebrates. Copper concentrations of 0.1 ppm are acutely toxic to nereis.[445] Perhaps most important is the sensitivity of some species of phytoplankton, whose photosynthesis can be inhibited by as little as 0.006 ppm copper.[12,22,155,212,228,334,393] In certain species of algae, copper appears to be as toxic as mercury (personal communication, L. Kamp-Nielsen).

The cupric ion appears to be the toxic agent for marine invertebrates, and fish may be protected from copper toxicity by chelating agents such as EDTA and nitrilotriacetic acid (NTA).[445,502] To protect fish, NTA must be used in a molar concentration at least three times that of the copper, and EDTA must be at least six times the amount of the element. Since these agents are quickly biodegradable in natural waters, they would be used chiefly as temporary treatments in unusual circumstances. For example, chelating agents could render a slug of toxic copper innocuous as it passed through a critical section of river.

The effectiveness of copper as a molluscicide to control schistosomiasis depends upon the lethal dose (LD). The LD is the product of concentration and time, usually expressed as parts per million-hours (ppm-h). As the time of exposure increases, the lethal concentration of the compound decreases disproportionately, so that the lethal dose also decreases. For example, the LD_{90} is 80 ppm-h when an organism is exposed to copper sulfate for 1 h, and only 14 ppm-h when exposed for 24 h.

Various copper compounds have been used as molluscicides since the 1920s, when copper sulfate was introduced for this purpose. Other compounds include copper pentachlorophenate, copper tartrate, copper carbonate, copper ricinoleate, copper resinate, cuprous oxide, copper dimethyldithiocarbamate, cuprous chloride, copper 3-phenylsalicylate, copper (II) acetylacetonate, and copper acetoarsenite (Paris green).[417] Most of these compounds are effective in two quite different ranges of concentration, depending on whether the treated water is clear—when they are effective at concentrations as low as 1 ppm—or turbid or muddy. In the latter case, mud seems to bind the copper or its compounds and reduce its toxicity to the snail, except when the animal actually ingests the copper–mud complex.

Attempts have been made to erect a "chemical barrier" molluscicide by continuously injecting very small concentrations of copper (0.1–0.3 ppm) into the river or lake.[136,155] Results have been varied and unpredictable.

Although other molluscicides have been developed to control schistosomiasis, copper—either in the ionic form or in one of the organic compounds listed above—remains a primary agent for this purpose.[326]

5

Human Copper Metabolism

DIETARY SOURCES

Copper is essential and generally considered a trace element, and almost every diet supplies a relatively large amount of the metal for body needs.[473] Indeed, it is difficult to prepare an acceptable diet that contains less than 2 mg of copper daily, and a single day's diet may contain 10 mg or more. The copper content of selected foods and beverages is listed in Table 5-1.

Oysters, liver, mushrooms, nuts, and chocolate are particularly rich in copper, but their copper content, like that of other foods, varies with the copper content of the soil or water in which they or their animal origins are grown.[572] Consequently, quantitative data differ widely from one tabulation to another. For example, one source indicates that kidney beans have almost no copper per 100 g edible portion,[141] whereas another reports a copper content of 0.11 mg/100 g.[318] The copper content of raw lobster in one tabulation is given as 2.2 mg/100 g, yet the next entry indicates that canned lobster has less than 1 mg/100 g.[141]

PATHWAYS FOLLOWED BY DIETARY COPPER

Studies on man using radioactive copper-64 (physical half-life, 12.8 h) or copper-67 (half-life, 61.8 h) indicate that about half of dietary copper

29

TABLE 5-1   Copper Content of Selected Foods and Beverages[a]

Food	Copper, mg/100 g edible portion[b]	Food	Copper, mg/100 g edible portion[b]
Fruits, Fruit Juices		Cauliflower, fresh	0.14
Apples, sweet, fresh	0.08	Corn (sweet), fresh	0.06
Apricots, fresh	0.12	Cucumbers, fresh	0.06
Dried	0.4	Dandelion greens, fresh	0.15
Avocados, fresh	0 .4	Eggplant, fresh	0.08
Bananas, fresh	0.2	Kale, fresh	0.09
Blackberries, fresh	0.12	Kohlrabi tubers	0.14
Blueberries, fresh	0.11	Lentils, dried	0.7
Cantaloupe, fresh	0.04	Lettuce, fresh	0.07
Cherries, fresh	0.07	Onions, fresh	0.13
Cranberries, fresh	0.09	Parsley, fresh	0.21
Currants (red), fresh	0.12	Parsnips, fresh	0.10
Dates, dried	0.21	Peas, green, fresh, unripe	0.23
Figs, fresh	0.06	Dried, split	0.80
Dried	0.4	Peppers (green), fresh	0.11
Gooseberries, fresh	0.08	Potatoes, raw	0.16
Grapes, fresh	0.1	Pumpkins, fresh	0.08 (0.03)
Grape juice	0.02	Radishes, fresh	0.13
Grapefruit, fresh	0.02	Rhubarb, fresh	0.05
Lemons, fresh	0.26 (0.04)	Soybeans, dried	0.11
Olives, green	0.46	Spinach, fresh	0.20
Oranges, fresh	0.07	Sweet potatoes, fresh	0.15
Orange juice, fresh	0.08	Tomatoes, fresh	0.10
Peaches, fresh	0.01	Turnips, fresh	0.07
Dried	0.3	Greens	0.09
Pears, fresh	0.13	Watercress	0.04
Pineapple, fresh	0.07		
Plums, fresh	0.3	*Nuts*	
Raisins, dried	0.2		
Raspberries, fresh	0.13	Almonds, dried	0.14
Strawberries, fresh	0.13 (0.02)	Brazil	1.1
Tangerines	0.1	Chestnuts, fresh	0.06
Watermelon	0.07	Coconuts, fresh	0.32
		Hazelnuts	1.35
Vegetables		Peanuts, roasted	0.27
		Pecans	—
Asparagus, fresh	0.14	Walnuts	0.31
Beans			
Kidney, fresh	— (0.11)	*Cereals, Cereal products*	
Lima, fresh	0.86		
Beets (beet roots, peeled, fresh)	0.19	Barley, pearled	0.4 (0.12)
		Cornflakes	0.17
Cabbage (red or white), fresh	0.06	Flour, buckwheat	0.7
Carrots, fresh	0.08	Rye, dark	—
		Wheat, whole	0.2

TABLE 5-1 *Continued*

Food	Copper, mg/100 g edible portion[b]	Food	Copper, mg/100 g edible portion[b]
White, unenriched	—	Bacon, fat, salted	—
Oatflakes	0.74	Beef, brain	0.2
Rice, polished, raw	0.06–0.19	Kidneys	0.35
Wheat germ	1.3	Liver	2.1
		Calf, liver	4.4
Sugar		Duck, medium fat	0.4
Dextrose, anhydrous	—	Goose, medium fat	0.3
Honey	0.2	Pork, loin or chops, cooked	— (0.09)
Sugar cane or beet, white	—	Turkey, medium fat	0.2
Fats		*Fish, Seafood* (raw unless otherwise stated)	
Lard	0.02		
Olive oil	0.07	Cod	0.5 (0.10)
		Flounder	0.18
Dairy products, Eggs		Haddock	0.23
Eggs, whole, raw	0.03	Halibut	0.23
Egg yolk, raw	0.02	Lobster	2.2
Milk (cow's) pasteurized, whole	0.01	Canned	—
Nonfat	—	Mackerel	0.16
Human breast milk	0.05	Oysters	3.6
Goat's milk	0.04	Salmon (Atlantic)	0.2
		Scallops	—
Meat, Poultry (raw unless otherwise stated)		Shrimp	0.4

[a] Derived from *Geigy Scientific Tables*[141] and *Low-Copper Diet*.[318]
[b] In instances where copper content given by *Geigy Scientific Tables* differs by 50% or more from the value supplied by *Low-Copper Diet*, the value from *Low-Copper Diet* is included in parentheses.

is not absorbed, but excreted directly in the feces.[512,530] When the amounts of separately administered oral and intravenous 2 mg doses of radioactive copper incorporated into the plasma copper protein, ceruloplasmin, were compared, the average absorption in 49 normal subjects was found to be 40%.[512] In another study, average absorption was 56% (40–70%) when the retention of 0.4–4.5 mg orally administered copper-64 was compared to the retention of simultaneously injected copper-67.[530] When 100 mg nonradioactive copper was administered as an emetic to children, about 30 mg was absorbed,[245] suggesting that the fraction absorbed from an oral dose of copper decreases little as the dose increases.

The form in and mechanism by which copper is absorbed and transported by the intestine are unknown. After entering the epithelial cells, it is taken up by a cytosol protein[158,507] similar to the metallothionein[156] of liver. Recent studies of copper absorption in infants suffering from an inherited, X-linked defect (Menkes's disease, discussed below)[118] suggest that the transfer of copper from intestinal cells to plasma involves an active process.[117]

Impaired copper absorption occurs in severe, diffuse diseases of the small bowel, including sprue, lymphosarcoma, and scleroderma.[78,516] Copper deficiency in plasma generally follows, but is of little, if any, clinical significance in the face of the multiple nutritional abnormalities symptomatic of these illnesses. The low serum concentrations of ceruloplasmin and copper are easily corrected with successful treatment of the underlying disease.

Immediately after it is absorbed, copper is transported in plasma bound to albumin[26] and perhaps to amino acids.[389,466] Almost all of the copper is soon deposited in the liver,[402,406,517] where about 80% of it is found in the cytosol bound to three proteins—hepatocuprein,[335] copper-chelatin or L-6-D,[369] and metallothionein (L-6-D).[156] The remaining 20% is incorporated into other specific copper proteins like cytochrome c oxidase, or is sequestered by lysosomes.[200,526]

Despite continued deposits of dietary copper, the hepatic concentration of the metal does not increase with age in man. The secretion of some copper into the blood following its incorporation into ceruloplasmin during hepatic synthesis[517] and the excretion of copper from the lysosomes into the bile[526] maintain this constant concentration. About 150–300 mg of ceruloplasmin containing 0.5–1.0 mg of copper is catabolized daily in the adult's liver[518] and about 30 mg ceruloplasmin, containing 0.1 mg copper, is excreted into the intestine.[601] Biliary copper may not be available for reabsorption because of its binding to a protein.[202] The rest of fecal copper comes from copper in salivary, gastric, pancreatic, and jejunal secretions,[474] along with any portion of dietary copper that has not been absorbed. Table 5-2 sets forth the copper content of various human tissues and body fluids.

COPPER PROTEINS

A protein containing copper may be a specific copper protein, such as ceruloplasmin; or, like albumin, it may bind the metal more loosely.[57]

That a protein is a specific copper protein requires two proofs. First, progressive purification must result in a ratio of copper to protein in which copper increases to an asymptotic value corresponding to an integral number of copper atoms per molecule of protein. Second, some properties of the metalloprotein should disappear when copper is removed, and reappear when a reversible recombination of copper and protein is effected. By applying these criteria, butyryl coenzyme A dehydrogenase, δ-aminolevulinic acid dehydrase, and β-mercaptotrans-sulfurase, all originally described as copper proteins,[253,298,324] subsequently have been shown only to be contaminated with the metal.[527,585,619]

Twenty mammalian copper proteins (Table 5-3) have been isolated but at least three of them—erythrocuprein, hepatocuprein, and cerebrocuprein—are identical,[81] and several have more than one name.

Albocuprein I and II   Albocuprein I and II are two pale yellow proteins recently isolated from human brains.[186] Their molecular weights are 14,000 and 72,000. Neither has any detectable enzymatic activity nor close similarities to cytocuprein (superoxide dismutase) or ceruloplasmin. Albocuprein I contains 0.25% copper, and albocuprein II holds 1.4% copper in its molecule. Both proteins also contain hexoses. Albocuprein II may be the primary copper-containing protein of the brain. The relation of these proteins to cerebral pathology has not yet been investigated.

Ceruloplasmin   Ceruloplasmin (polyamine oxidase, ferroxidase I; EC 1.12.3.1) is a blue plasma glycoprotein containing 0.3% copper and 8% carbohydrate.[61-63,367,472,489,490] Depending on the analytic method employed, its molecular weight is between 132,000[323,460] and 160,000.[268] To a moderate degree, it catalyzes the oxidation of several polyamines, catecholamines, and polyphenols, particularly p-phenylenediamine (PPD), and Fe (II) to Fe (III).[113,242,268,405,433,460]

The last reaction may be essential in some species for the uptake of ferric iron by apotransferrin. Recently histaminase activity has also been attributed to ceruloplasmin.[221] Except for ferroxidase activity,[456] all of these enzymatic activities have been observed only *in vitro,* and their physiologic significance is obscure. The purpose of the reversible dissociation of the molecule's copper is similarly enigmatic,[368] although small amounts of apoceruloplasmin have been reported to be present in human[83] and rat serum.[244]

The 8-10 oligosaccharide chains of ceruloplasmin are composed principally of glucosamine, mannose, and galactose; almost all—if not

34

COPPER

TABLE 5-2   Copper Content of Human Tissues and Body Fluids[a]

Tissues	μg/g dry weight[b]		mg copper/whole organ or tissue[c]	
	Mean	Range	Median	80% Range
Adrenal	7.4	1.1–28.9	0.02	0.01–0.02
Aorta	6.7	2.4–21.9	–	–
Bone	4.2	0.9–11.8	–	–
Brain	23.9	13.1–39.4	8.1	5.4–11.3
Breast	4.6	1.4–8.4	–	–
Erythrocytes (per 100 ml packed red blood cells)	89.1	63.0–107.0	–	–
Hair	23.1	7.4–54.5	–	–
Heart	16.5	10.1–22.9	1.2	0.8–1.5
Kidney	14.9	5.1–35.7	0.9	0.8–1.1
Leukocytes (per 10⁹ cells)	0.9	0–1.4	–	–
Liver	25.5	9.2–46.8	11.3	7.1–28.7
Lung	9.5	4.2–15.9	1.3	1.0–2.0
Muscle	5.4	2.0–13.8	26.7	18.0–43.2
Nails	18.1	3.2–58.2	–	–
Ovary	8.1	3.1–16.5	0.009	0.007–0.01
Pancreas	7.4	2.4–20.0	0.1	0.08–0.2
Placenta[364]	13.5	11.8–16.6	–	–
Prostate	6.5	1.8–11.0	0.02	0.01–0.03
Skin	2.0	0.3–5.4	1.4	1.1–2.1
Spleen	6.8	3.1–16.1	0.2	0.1–0.3
Stomach and intestines	12.6	4.5–36.6	3.3	1.1–2.6
Thymus	6.7	3.3–11.5	–	–
Thyroid	6.1	1.6–17.5	0.02	0.009–0.05
Uterus	8.4	3.5–25.2	–	–

all—chains are terminated by a sialic acid residue. The presence of this residue appears to be essential to the survival of the protein in the circulation.[213,366]

Copper-chelatin (L-6-D)   Copper-chelatin, a cytoplasmic copper-binding protein of about 8,000 daltons, has been isolated from livers of rats.[621] It is characterized by a high sulfydryl concentration and a content of six atoms of copper per molecule. In the human liver, a similar

TABLE 5-2 *Continued*

	μg/100 ml	
Body Fluids	Mean	Range
Aqueous humor[190]	12.4	—
Bile[474]	—	24.0–538.0
Cerebrospinal fluid[266]	27.8	10.0–70.0
Gastric juice[474]	28.1	0.0–200.0
Pancreatic juice[474]	28.4	0.0–69.0
Plasma, Wilson's disease[86]	50.0	33.0–65.0
Saliva[132]	31.7	5.0–76.0
Serum, female[87]	120.0	87.0–153.0
Serum, male[87]	109.0	81.0–137.0
Serum, newborn[470]	36.0	12.0–67.0
Sweat, female[240]	148.0	59.0–228.0
Sweat, male[240]	55.0	3.0–144.0
Synovial fluid[392]	21.0	4.0–64.0
Urine (24 h)[77]	18.0	3.9–29.6

[a] Derived from Scheinberg and Sternlieb.[474]
[b] Values for most tissues in these two columns derived from Fell *et al.*,[167] except where other references are indicated.
[c] Values in these two columns are from Sass-Kortsak.[467]

protein (L-6-D)[369] is found in concentrations of about 5 mg/g wet tissue, although its copper content may vary.

Cytochrome c oxidase Cytochrome *c* oxidase (EC 1.9.3.1), present in the mitochrondria of many animal and plant tissues, contains one heme molecule, one iron atom, and apparently two copper atoms per molecule, and weighs about 270,000 daltons. Studies of the paramagnetic resonance and oxidation–reduction characteristics of beef heart cytochrome oxidase indicate that its copper is of two different species.[154,374,631] Copper is essential to the protein's enzymatic activity:[380] severe copper deficiency, whether acquired or hereditary, is generally associated with reduced cytochrome oxidase activity.[115,177,178]

Dopamine β-hydroxylase Dopamine β-hydroxylase (3,4-dihydroxy-phenylethylamine β-hydroxylase; EC 1.14.2.1) is an oxidase that weighs about 290,000 daltons and contains 4-7 copper atoms per molecule.[182] It catalyzes the conversion of dopamine to norepinephrine.[501] In preparations isolated from beef adrenal glands, about one-third of the copper

TABLE 5-3   Mammalian Copper Proteins

| Protein | Isolated from | |
	Species	Organ or Tissue
Albocuprein I	Man	Brain
Albocuprein II	Man	Brain
Ceruloplasmin	Numerous, including man	Plasma
Copper-chelatin (L-6-D)	Man, rat	Liver
Cytochrome c oxidase	Numerous	Heart, liver, etc.
3,4-dihydroxyphenylethylamine β-hydroxylase	Cattle	Adrenals
Dopamine β-hydroxylase	Cattle	Adrenals
Ferroxidase II	Man	Serum
Hepatomitochondrocuprein	Man, cattle	Liver
Lysyl oxidase	Chicken	Cartilage
Metallothionein	Man, cattle	Liver
Mitochondrial monoamine oxidase	Man, rat, cattle	Liver, brain
Pink copper protein	Man	Erythrocytes
Plasma/serum monoamine oxidase	Man, rabbit, pig	Plasma/serum
Superoxide dismutase (cytocuprein)		
Cerebrocuprein	Man	Brain
Erythrocuprein	Man	Erythrocytes
Hemocuprein	Man	Blood
Hepatocuprein	Man	Liver
Tryptophan-2,3-dioxygenase	Rat	Liver
Tyrosinase	Man	Skin, eye

atoms are cuprous and the enzymatic hydroxylation appears to require reduction of about another third to the cuprous state. Hydroxylation of the substrate is accompanied by a stoichiometrically equivalent oxidation of one-third of the enzyme's cuprous atoms to cupric.[182] The enzyme is also found in cattle brains and hearts.

Ferroxidase II   Ferroxidase II, thought to be another copper protein that can catalyze the oxidation of ferrous iron, has been isolated from the serum of normal individuals and of patients with Wilson's disease. Ferroxidase II is yellow, contains 0.1% copper, and displays no oxidase activity toward PPD.[567]

Hepatomitochondrocuprein   Hepatomitochondrocuprein contains about 3% copper and has been isolated from crude mitochondrial fractions of bovine and neonatal human liver.[434,435] Despite its name, the actual subcellular locus of this protein is probably lysosomal.[436]

Lysyl oxidase Lysyl oxidase, an enzyme that has been isolated from chick cartilage, converts certain lysyl residues in collagen and elastin into the corresponding adipic semialdehydes. The enzyme is almost certainly a copper protein: Extracts of cartilage from chicks raised on a copper-deficient diet have no lysyl oxidase activity. Dialysis of the purified enzyme against the chelating agent α,α'-dipyridyl abolishes its enzymatic activity; activity is restored when the apoenzyme is dialyzed against 0.001 M copper chloride. Deficient activity of this enzyme may be related to the structural abnormalities seen in collagen, elastin, and keratin in natural and experimental copper deficiencies.[491]

Metallothionein Metallothionein is a colorless protein, first described by Vallee and co-workers;[583] it has a molecular weight of about 10,000 daltons, a high sulfhydryl content, and the capacity for binding copper, zinc, and cadmium. It has since been isolated from the cytosol of human,[157] rat, chicken,[613] cow,[56] and horse liver;[265] equine renal cortex;[339] and chick, cow, and rat duodenal mucosa.[158] The equine liver and kidney protein appears to have a monomeric molecular weight of 6,000 daltons.[265]

Mitochondrial monoamine oxidase Mitochondrial monoamine oxidase (EC 1.4.3.4), present in the mitochondria of liver and brain, is enzymatically active toward the same substrates as the plasma enzyme, and it also oxidizes epinephrine and serotonin. Copper is essential for its enzymatic activity. Mitochondrial monoamine oxidase has not been solubilized.[385,395]

Plasma/serum monoamine oxidase Plasma/serum monoamine oxidase (EC 1.4.3.4) is a copper protein weighing 255,000 daltons. It has been purified from human and rabbit serum, and from steer and hog plasma.[351,628] *In vitro,* the enzyme catalyzes the oxidative deamination of several monoamines to form hydrogen peroxide, ammonia, and the corresponding aldehydes.[352] Its physiologic and pathologic importance is obscure, although its concentration increases in the presence of congestive heart failure or parenchymal liver disease.[353]

Superoxide dismutase Superoxide dismutase (cytocuprein) is also called erythrocuprein (or hemocuprein), hepatocuprein, or cerebrocuprein, according to whether it has been isolated from the cytosol of erythrocytes, liver, or brain.[81,181] A light-blue-green metalloprotein with a molecular weight of 33,600 daltons, it catalyzes the dismutation of the superoxide anion into hydrogen peroxide and oxygen[284,349,487,506] and may also scavenge oxygen atoms in certain reactions. For optimal enzymatic activity, at least two of the protein's four metal ions must be cupric.[82, 259, 612] About 30 mg of the protein are present in 100 ml of packed normal human erythrocytes, accounting for more than 60% of total erythrocyte copper.[340] This protein—or a very similar one—is also present in human kidney, thyroid, pituitary, and adrenal glands.[227] Recently a protein containing one copper atom and exhibiting superoxide dismutase activity was isolated from the marine bacterium *Photobacterium leiognathi.*[441]

Tryptophan-2,3-dioxygenase   Tryptophan-2,3-dioxygenase (EC 1.13.1. 12) has been isolated from rat liver cytosol.[482] It is a heme protein with a molecular weight of 167,000 daltons and two cuprous atoms per molecule. This enzyme catalyzes the insertion of molecular oxygen into the pyrrole ring of L-tryptophan.[53]

Tyrosinase   Tyrosinase, found principally in the melanocytes of skin and eye, contains about 0.2% copper.[65] Actually, it may be a series of similar proteins, each of which catalyzes one of the sequential reactions that convert tyrosine to melanin.[172-174] A genetic abnormality in tyrosinase may cause the tyrosinase-negative, oculocutaneous form of albinism inherited as an autosomal recessive trait.[171]

A pink copper protein, still unnamed, has been isolated from erythrocytes. Although its molecular weight is close to that of superoxide dismutase, it differs from the latter in spectral, electrophoretic, chromatographic, and enzymatic characteristics.[448] Its biologic role is unknown.

COPPER DEFICIENCY

Copper deficiency is a rare condition, because the normal infant is born with a generous store of copper in its liver.[373, 483, 618] However, Josephs suggested in 1931 that copper deficiency might account for a resistant anemia in milk-fed infants,[263] because cow's milk is one of the few major foods deficient in copper (Table 5-1). Indeed, during the past several years, numerous instances of copper deficiency associated with anemia, neutropenia, and severe demineralization of the bone in both premature and full-term infants have been reported.[9, 13, 102, 103, 209, 484] Induction of clinically significant copper deficiency seems to require that either severe malnutrition or intestinal malabsorption be present. Fortification of the feeding formula with iron may impair the absorptive mechanisms for copper and lead to copper deficiency.[484] Copper deficiency has been observed in an infant with ileal atresia[267] and in adults maintained on parenteral nutrition for prolonged periods.[591] Additions to the diet or infusions of 1 mg copper/day (except for the premature infant, who should receive only 100 $\mu g/kg$)[101] is more than adequate supplementation in all these instances, and is capable of reversing the anemia, neutropenia, and bone lesions that are the most significant clinical effects of copper deficiency in infants.[102,103]

Menkes's Disease (Trichopoliodystrophy)

Steely- or kinky-hair disease, first described as a syndrome by Menkes *et al.*[359] in 1962, is an X-linked fatal disorder. Affected male infants exhibit kinky, depigmented hair; physical and mental retardation, with widespread degeneration of the brain; hypothermia; and death within the first several years of life.[359]

In 1972 Danks *et al.*[118] demonstrated copper deficiency in infants with markedly low serum concentrations of copper and ceruloplasmin. They pointed out that the kinky or steely depigmented hair was similar to the abnormal wool seen in copper-deficient sheep, and believed to be caused by defective cross-linking of keratin. Similar defective cross-linking in collagen and elastic—probably a consequence of abnormally low concentrations of the copper-containing enzyme, lysyl oxidase— also was thought to lead to degeneration of the internal elastic lamina of arterial walls, in turn causing brain degeneration.[118] Brain degeneration may also be related to a deficiency of cytochrome *c* oxidase.[177,178] A generalized defect in transport of copper across membranes, rather than defective absorption from the gastrointestinal tract alone, may underlie the disorder.[117]

Unfortunately, therapeutic administration of copper, whether oral[118] or parenteral,[214] has done little more than raise concentrations of ceruloplasmin and copper in serum. No definite clinical improvement or decrease of mortality has occurred.

An X-linked inherited defect in mottled mice is associated with pathologic findings somewhat similar to Menkes's syndrome; it is also associated with defective copper transport and deficiency.[252] These abnormal mice and the affected Bedlington terriers mentioned below may provide animal models of two serious inherited disorders of human copper metabolism.

In adults, copper deficiency can occur when intake or absorption of the metal is drastically reduced,[78,102,516] when disease causes prolonged, excessive urinary or intestinal loss of ceruloplasmin,[447,601] or when the copper balance becomes negative during the prolonged administration of D-penicillamine (β-mercaptovaline) to individuals with normal copper stores.[231] All of these conditions are associated with lowered blood concentrations of ceruloplasmin and copper.

Occasionally, administration of D-penicillamine may cause copper deficiency accompanied by impaired taste acuity that will return to normal if a few milligrams of copper salt are added to the daily diet.[231]

TOXICITY

A survey of 969 water systems located in nine geographic areas in the United States showed average copper concentrations of 134 μg/l.[347] The highest concentration was 8,350 μg/l, and 1.6% of the 2,595 samples exceeded the drinking water standard of 1 mg/l. This limit on copper in drinking water was established not because toxicosis was of concern, but because excessive concentrations impart an undesirable taste to water. Higher concentrations are more frequent in acidic, soft waters; at a pH below 7.0, 5% of the samples exceeded the standard. The concentration of copper in drinking water is rarely high enough to affect its taste or produce toxicosis.

Acute Toxicosis

Acute copper poisoning occurs in man when at least several grams of copper sulfate are ingested or when acidic food or drink—vinegar, carbonated beverages, or citrus juices—have had prolonged contact with the metal.[42, 246, 355, 410, 485] When carbonated water remains in copper check-valves of drink-dispensing machines overnight, the copper content of the first drink of the day may be increased enough to cause metallic taste, ptyalism, nausea, vomiting and epigastric burning, and diarrhea.[246] Whiskey sours and fruit punches mixed or stored in copper-lined cocktail shakers or vessels have had the same effects.[2, 355] A whiskey sour that contained 120–135 ppm copper, or approximately 10 mg of cupric ions in a 60–90 ml drink, produced abdominal cramps, vomiting, and diarrhea within 10–90 min of ingestion.[2] Eight children and one of two adults who drank an orange-flavored beverage refrigerated overnight in a brass pot became nauseated, and several of the children vomited. The drink contained 34 ppm copper, so that 240 ml would have supplied 8.5 mg copper; but since the children were aged 1–4, it is likely that smaller amounts of the beverage and less copper were ingested.[1] In eight other instances of acute copper poisoning involving over two dozen individuals and reported by the Center for Disease Control since 1968,[576] copper plumbing or vessels, with which acidic (generally carbonated) water had prolonged contact, led to the toxicosis.[577-579]

The vomiting and diarrhea induced by ingesting milligram quantities of ionic copper generally protect the patient from its serious systemic toxic effects: hemolysis, hepatic necrosis, gastrointestinal bleeding, oliguria, azotemia, hemoglobinuria, hematuria, proteinuria, hypotension, tachycardia, convulsions, coma, or death.[95,119] When more than gram quantities of a salt such as copper sulfate are ingested—generally with suicidal intent—gastrointestinal mucosal ulcerations, hemolysis, hepatic necrosis, and renal damage from deposition of hemoglobin and/or copper constitute the pathogenetic factors underlying these effects.[95,600]

Hemolysis has also been reported after applying solutions of copper salts to large areas of burned skin,[243] or after introducing copper into the circulation during hemodialysis.[320] The source of this copper may come from the semipermeable membranes (generally fabricated with copper) and copper tubing or heating coils of the dialysis equipment. Copper in the membrane appears to be transferred to the patient; in one instance a Cuprophan membrane introduced 632 μg copper into a patient.[1] Copper from tubing or coils seems hazardous only when the dialysate becomes acidic; the pH of the circulating fluid can drop to 2.5 when a deionizer in the circuit is exhausted.[18,39,40,286,336] Therefore, even 1-2 yr of twice weekly dialysis can raise hepatic copper concentration to abnormally high levels.[18,39] Copper introduced into the circulation by hemodialysis can produce febrile reactions remarkably similar to classical metal fume fever experienced by workers in copper smelters and refineries.[320] Similarly, copper stopcocks in circuits used for exchange transfusions have been reported as the source of potentially hazardous infusions of copper for neonates.[38]

Chronic Toxicosis

Presence of a fragment of metallic copper in the eye (chalcosis bulbi) may result in loss of the eye, sunflower cataracts, or visible deposits of copper in the cornea known as Kayser–Fleischer rings.[223,457] Prolonged administration of D-penicillamine may be the only available nonsurgical therapy.

Bordeaux mixture, a 1-2% solution of copper sulfate neutralized with hydrated lime, is used widely to prevent mildew on grapevines, particularly in France, Portugal, and southern Italy. Pulmonary copper deposition and fibrosis occur in the lungs of some vineyard workers after years of exposure to such solutions.[429] Their lungs may be blue, suggesting the presence of excess copper. More recently, granulomas and malignant tumors have appeared in these laborers' livers and lungs.[430,590] In contrast, studies of Chilean copper miners show that liver and serum concentrations of copper are normal,[474] despite years of exposure to copper sulfide and oxide dusts, which are insoluble.

Drinking water with an unusually high copper concentration (800 μg/l) may have caused acrodynia (pink disease) in a 15-month-old infant.[462] Gingivitis,[568] lichen planus,[184] and eczematous dermatitis[21] have been attributed to the copper alloys used in some dental and other prostheses. Wearing copper bracelets as apocryphal treatment for arthralgias only leads to green-stained skin.

From the rarity with which human copper deficiency or toxicosis occurs, it appears that humans are able to avoid both despite wide variations in dietary supply.

Wilson's Disease (Hepatolenticular or Hepatocerebral Degeneration)

Genetics Wilson's disease has been found in every racial group in which it has been sought.[25,289,476,607,620] The illness is inherited as an autosomal recessive trait with a general prevalence of about 1 in 200,000. This figure is consonant with the possession of 1 "Wilson's disease gene" by 1 in about 200 people.* The heterozygotes remain free of pathologic manifestations of Wilson's disease. As with any recessive disorder, the incidence will be higher in locales where a greater amount of inbreeding occurs.

Pathogenesis and pathology Almost all patients with Wilson's disease exhibit a lifelong deficiency of the plasma copper-protein ceruloplasmin,[471] and an excess of hepatic copper.[200,476,511,522,524] This excess copper in the liver may be caused, in part, by impairment of lysosomal excretion of hepatic copper into bile,[183,403,407,526,531] and is associated with diminished or absent hepatic synthesis of ceruloplasmin.[518] It is remarkable that retaining only 1% of the dietary intake of copper (10–20 mg/yr) is sufficient to cause Wilson's disease. Table 5-4 lists concentrations of copper found in tissues of untreated patients with this disorder.

During the early stages of Wilson's disease, the liver is capable of binding as much as 30–50 times its normal concentration of copper with little, if any, overt clinical disorder.[522] Ultimately, hepatic copper is

---

*Since 1 in 40,000 nonconsanguineous marriages will be between such heterozygotes, and since one-quarter of their children will inherit a pair of abnormal alleles, Wilson's disease will develop in about 1 in 160,000 people.

released into the bloodstream, and in the face of massive necrosis of hepatic parenchyma, that action may suddenly infuse large amounts of copper into the plasma, inducing severe hemolysis and jaundice.[455,511,524] In most patients, however, the metal diffuses into the circulation gradually, causing the plasma concentration of free copper to rise 5–10 times, to about 25–50 μg/100 ml. This copper diffuses out of the vascular compartment into extracellular fluids and tissues with toxic effects in susceptible cells. Characteristic ultrastructural changes, fatty degeneration of hepatocytes eventuating in necrosis, collapse of parenchyma, and postnecrotic cirrhosis occur in the liver.[166,511,524] Later, the excess hepatocellular copper is sequestered by lysosomes, a process that seems to render the metal innocuous for other cytoplasmic organelles.

Unless the patient with Wilson's disease succumbs to hepatic necrosis, the toxic effects of copper ultimately are manifested primarily in the central nervous system and kidneys. In the corneas, copper deposits are visible as pathognomonic Kayser–Fleischer rings or crescents[25,108,192,289,398,452,477,481,624] best seen with the slit lamp as golden or greenish-brown pigment grains in the periphery of Descement's membrane.[514,536] In some patients, copper is also deposited on the capsular surfaces of the lens as a sunflower cataract.[79]

Diagnosis In about half of all patients, the first clinical evidence of Wilson's disease represents dysfunction of the liver. Ascites, esophageal variceal hemorrhage,[525] a syndrome mimicking toxic or infectious hepatitis, hemolysis caused by sudden release of sequestered copper,[130] deficiency of clotting components, hypersplenism, or gonadal dysfunction may be manifested.[475,476,524]

In almost all other patients, neurologic or psychiatric disorders are the initial clinical manifestation. The neurologic picture may resemble parkinsonism, multiple sclerosis, chorea, dystonia, or any combination of these diseases; or it may be *sui generis*.[25,134,289,476,607] The usual onset is insidious. Dysarthria is a frequent sign in children, and often subtle incoordination, resting or intentional tremors, athetoid movements, rigidity, or dystonic posturing and distortion can occur at all ages. Excessive salivation and drooling are often troublesome. Epileptiform seizures have been reported but are unusual.[414,495] Specific disturbances in reflexes, sensation, or muscular strength are so rare in Wilson's disease that their presence should lead the physician to search for another diagnosis.

Psychiatric disorders may accompany the neurologic symptoms or precede any other evidence of disease.[24,201,299] In young adults, the spectrum ranges from mild behavioral disturbance difficult to differen-

TABLE 5-4 Copper Content of Tissues from Patients with Untreated Wilson's Disease (µg/g dry tissue)

Tissue	Cartwright et al.[86]	Tu et al.[a]	Leu et al.[311]	Bauman[23]	Bickel et al.[a,35]	Cumings[108]	Wisniewski et al.[622]
Adrenal	(1)[b] 9.6	(2)[b] 3.6 [0.9]	(1)[b] 64.0	(1)[b] 3.0	(2)[b] 8.0	(4)[b] —	(4)[b] —
Caudate nucleus	—	—	—	—	—	189.0	235.0
Cerebellum (white matter)	—	—	—	—	—	247.0	275.0
Cornea	12.0	64.4 [16.1]	—	11.0	284.0	—	—
Frontal lobe, cortex	—	—	—	—	—	135.0	101.0
Globus pallidus	—	—	—	—	—	220.0	289.0
Heart	12.0	10.8 [2.7]	—	11.0	17.0	—	—
Jejunum	—	5.6	—	26.0	10.0	—	—

Kidney	47.2	125.6 [31.4]	—	46.0	166.0	—	—
Liver	532.0	579.6 144.9	792.0	—	433.0	—	—
Lung	15.2	3.6 [0.9]	26.0	—	17.0	—	—
Muscle	10.0	4.3 [1.1]	102.0	8.0	5.0	—	—
Ovary	—	5.2 [1.3]	—	—	—	—	—
Pancreas	4.2	—	—	—	—	—	—
Putamen	—	—	—	—	—	362.0	265.0
Skin	—	5.2 [1.3]	—	—	—	—	—
Spleen	7.7	4.0 [1.0]	—	1.5	9.0	—	—
Stomach	—	—	63.0	—	10.0	—	—

[a] Numbers represent means of two patients, except for those noted for heart and cornea; figures in brackets are $\mu g/g$ wet weight, as in Tu et al.[569]

[b] Parentheses denote number of patients studied.

tiate from the normal vicissitudes of adolescence, through marked deterioration of schoolwork and neurosis, to the manic-depressive or schizophrenia-like psychosis, which may appear in all age groups. The emotional disturbance may be partly a reaction to the somatic dysfunction. Yet it is hard not to conclude that cerebral deposits of copper (Table 5-4) must also exert a direct toxic effect on higher brain centers, although no specific psychiatric syndrome has been attributed to Wilson's disease as yet. However, very few sophisticated psychiatric studies of patients with this disorder have been made.

In rare instances, hematuria has been cited as the first evidence of the toxic effects of copper.[167]

Relatives, particularly siblings, of patients with Wilson's disease must be examined even if they appear perfectly healthy, because of the autosomal recessive transmission of the illness.[522] Fortunately, biochemical findings in patients with the manifest disease have made it possible to confirm the diagnosis in an asymptomatic individual when there are less than 20 mg ceruloplasmin/100 ml serum *and* more than 250 μg copper/g dry liver.[511]

Treatment   From Wilson's treatise in 1911 until 1948, this disease was considered progressive and fatal. The recognition of the etiologic role of copper and Cumings's suggestion that British antilewisite (BAL; 2,3-dimercapto-1-propanol) might halt that progression led to successful specific therapy.[109, 135] In 1957, an effective oral treatment to remove copper with penicillamine was introduced by Walshe,[605] and now it is generally recognized that specific treatment of the disease produces dramatic results.[131, 513, 521, 604, 607] Even prevention of all overt manifestations and an apparently normal life span are feasible for asymptomatic patients with Wilson's disease; they require only the regular, lifelong administration of penicillamine.[522] Recently Walshe introduced triethylene tetramine dihydrochloride as a new chelating agent for the treatment of Wilson's disease.[606]

Biochemical Basis of Copper Toxicosis

The effects of copper in organs, tissues, cells, and subcellular organelles from patients with Wilson's disease are compared here with data derived from *in vitro* studies and natural or experimental copper toxicoses in animals. The Dominican toad, *Bufo marinus*, normally accumulates concentrations of hepatic copper comparable to those seen in humans with Wilson's disease,[199] but copper is sequestered in the toad's lysosomes where, as in the newborn baby,[198] it does not seem to cause pathologic changes. Experimental copper poisoning of animals has clarified only a few biochemical, physiologic, and pathologic aspects of copper metabolism in the liver, brain, and kidney. Very recently, a form of chronic copper toxicosis (probably inherited) has been reported to occur in certain Bedlington terriers.[225]

Liver Chronic administration of copper to rabbits, rats, mice, pigs, and sheep forces deposition of the metal in the liver, as well as in other organs.[8, 19, 20, 142, 165, 197, 303, 313, 330, 362, 588, 623, 626] Copper accumulates when the excretory capacity of the liver cell is exceeded. Fractionation of liver homogenates (Table 5-5) and morphologic studies on such copper-loaded animals show:

• an increase in hepatic copper concentration, which is even more marked if there is bile duct obstruction;[626]
• a change in the relative subcellular distribution of copper with a 200–300% increase in the proportion in the mitochondria and lysosomes, and a marked decrease in the proportion in cytosol of hepatocytes;[165, 588] and
• increased numbers and prominence of copper-containing lysosomes.[20, 197]

This induction of lysosomes is not evident in children and adolescents with Wilson's disease, because their organelles are inconspicuous despite high cytoplasmic concentrations of the metal.[511, 522] In older patients, however, the numbers of lysosomes increase and the metal is sequestered as the disease progresses.[200] This phenomenon seems to protect the liver cells from the cytotoxic effects observed in the younger patients. In contrast, rats experimentally poisoned by copper release acid phosphatase and other hydrolases from lysosomes to cytosol and microsomes,[314, 356] which does not occur in copper-poisoned mice[588] or in patients with Wilson's disease.[526]

Human, rat, and cow hepatic cytosol contains two or three copper-binding proteins weighing about 8,000, 10,000, and 40,000 daltons.[41, 156, 369, 526, 621]

The lower molecular weight protein isolated from patients with Wilson's disease appears identical to that from control subjects.[369] Two recent studies concluded that metallothionein samples from Wilson's disease patients have a greater than normal affinity for copper.[157, 469]

TABLE 5-5   Experimental Biochemical, Physiologic, or Ultrastructural Effects of Copper on Liver

Study No.	Preparation	Species	Copper Added	Concentration of Copper, $\mu g/ml$	Effect
1[598]	Homogenate	Rat	in vivo	—	Glycogen content increased; phospholipid decreased
2[313]	Homogenate	Rat	in vivo	—	Acid phosphatase activity increased
3[208]	Mitochondria	Rat	in vitro	1.0	Respiration inhibited; oxidative phosphorylation uncoupled
4[587]	Mitochondria	Rat	in vitro	0.2	Swelling
5[20]	Mitochondria	Rat	in vivo	—	Cristae rearranged
6[588]	Mitochondria	Mouse	in vivo	—	Malate dehydrogenase activity decreased
7[365]	Microsomes	Rat	in vivo and in vitro	— 0.9	Aniline hydroxylase activity decreased
8[356]	Microsomes	Rat	in vivo	—	Acid phosphatase activity increased
9[232]	Microsomes	Rabbit	in vitro	10.0	Heavy polysomes diminished
10[20,232,588]	Lysosomes	Rat	in vivo	—	Lipofuscin granules more numerous
11[314]	Lysosomes	Rat	in vivo	—	Lysosomal membranes broken
12[314]	Lysosomes	Rat	in vitro	6.3–25.0	Acid phosphatase released
13[588]	Lysosomes	Mouse	in vivo	—	Acid phosphatase, β-glucuronidase, and N-acetyl-β-D-glucosaminidase activities increased
14[356]	Cytosol	Rat	in vivo	—	Acid phosphatase activity increased

Erythrocytes   In patients with Wilson's disease and in sheep with chronic copper toxicosis, massive or submassive necrosis of the liver sometimes can free large enough amounts of copper in a sufficiently short time to cause considerable hemolysis.[130,354,415,455,566,617] Since copper analyses are rarely performed in blood at the height of the hemolytic process, it is difficult to pinpoint the maximal concentration of free copper in the plasma, which is the probable cause of hemolysis. Estimation of this value is also unreliable, because free plasma copper is the difference between total copper and ceruloplasmin-bound copper, and usually both measurements are not given in case reports. In one of the published reports, the value of free plasma copper given during the hemolytic crisis was about 0.7 μg/ml,[130] or about 10 times the normal free copper concentration. In another patient, 1.8 μg/ml was present temporarily in the plasma at the peak of hemolysis. A qualitative estimate of the concentration of free plasma copper may be made from studies of daily urinary copper excretion. Excreted copper rarely exceeds 0.5 mg/day in nonhemolyzing, untreated Wilson's disease patients, because free plasma copper is the only source of urinary copper. In each of five patients not receiving penicillamine, the highest 24-h urinary excretion measured during hemolysis was 1.2,[566] 3.0,[130] 2.4, 2.5, and 4.7 mg.[354] Table 5-6 lists the results of studies showing biochemical effects of ionic copper on several intraerythrocytic compounds. The data show that only six of the enzymes studied by Boulard et al.[46] and the pyruvate kinase investigated by Willms et al.[617] were affected by the concentrations of free copper attained in patients. A hereditary defect in any one of these enzymes is generally associated with nonspherocytic hemolytic anemia; thus it is attractive to speculate that such copper inhibition *in vivo* is the direct cause of hemolysis. However, measurement of the activities of these six enzymes in erythrocytes from three patients with Wilson's disease during a hemolytic crisis, and from one patient two years after hemolysis, showed *increased* activity in every instance.[617] In another patient with Wilson's disease, normal values for glutathione reductase*, glucose-6-phosphate dehydrogenase, and pyruvate kinase were noted during and after the hemolytic crisis. Therefore, the cause of hemolysis in Wilson's disease remains unknown.

Brain   A number of studies on the effects of copper on the respiration of preparations of neural tissue have shown that either facilitation or inhibition results, depending on experimental conditions.[477] The data do not appreciably contribute to an understanding of the mechanisms underlying the neurophysiology or neuropathology of copper.[15, 48, 76, 94, 153, 423, 424, 497, 554, 595, 596]

---

*Values were above normal for four patients studied by Willms et al.[617] and not measured by Boulard et al.[46]

TABLE 5-6   Experimental Biochemical Effects of Copper
on Human Erythrocytes

Study No.	Preparation	Concentration of Copper, μg/ml	Effect
1[159]	Nicotinamide adenine dinucleotide phosphate, reduced (NADPH)	63.0	Oxidation enhanced
	Glucose-6-phosphate dehydrogenase (G-6-PD)	63.0	Inhibited
	Glutathione	63.0	Oxidized
2[415]	G-6-PD	6.3	Inhibited
3[130]	Glutathione	6.3–63.0	Content and stability diminished
	G-6-PD	6.3–63.0	Inhibited
	Glutathione reductase	6.3–63.0	Inhibited
	NADPH	6.3–63.0	Oxidation unchanged
4[360]	Glutathione	2.5	Content diminished
	Catalase	5.0	Diminished
	Glutathione reductase	5.0	Diminished
5[617]	Pyruvate kinase	0.3–1.89	Inhibited
6[164]	Adenosine triphosphate (ATP)	63.0	Utilization diminished
7[46]	Hexokinase	0.95–6.3	Inhibited
	Phosphofructokinase	0.95–6.3	Inhibited
	Phosphoglyceric kinase	0.95–6.3	Inhibited
	Pyruvate kinase	0.95–6.3	Inhibited
	6-Phosphogluconate dehydrogenase	0.95–6.3	Inhibited
	G-6-PD	3.15–6.3	Inhibited

Cholestasis

Copper frequently accumulates in the livers of patients with chronic
extra- or intrahepatic cholestasis.[250,494,627] Two- or threefold elevations
of hepatic copper concentration have been reported in these disorders.
More marked elevations (even as high as 1,200 μg/g dry weight), in the
range seen in patients with Wilson's disease, occur in children with
biliary atresia[515] and in some patients with biliary cirrhosis.[627] The
etiologic role of copper in these disorders certainly is not primary, but it
is possible that copper toxicosis may aggravate the severity of their
degenerative process. Nevertheless, a role for copper-chelating therapy
has not been established in the treatment of biliary cirrhosis.

ADDITIONAL CLINICAL ASPECTS OF
COPPER METABOLISM

Diagnostic Value of Measuring Serum Copper and
Ceruloplasmin Concentrations

The normal range of concentration of ceruloplasmin in healthy adults is
20-45 μg/100 ml of serum or plasma, corresponding to a copper concentration of about 80-140 μg/100 ml. Normally, 95% or more of serum
copper is integrally bound to ceruloplasmin; the rest is loosely bound to
albumin. Consequently, a measurement of the concentration of either
serum copper or ceruloplasmin usually serves as an accurate measure of
the other. The only important exception occurs in patients with Wilson's
disease, who may have little or no ceruloplasmin, yet exhibit appreciable
concentrations of serum copper (Table 5-2).

Serum ceruloplasmin is reduced in all normal neonates[432] and is
diagnostically lowered in patients with Wilson's disease[471,476] and
Menkes's syndrome.[118] There is little diagnostic significance in the
lowered concentrations found in association with severe malnutrition
and malabsorption syndromes,[516,601] massive hepatic necrosis,[608] the
nephrotic syndrome,[447] and protein-losing enteropathies.[601]

Late in pregnancy, the serum concentration of ceruloplasmin increases
2-3 times.[213] The administration of estrogens or their analogues also
brings on similar elevations.[84]

Increased concentrations of ceruloplasmin and copper are seen in
rheumatoid arthritis, rheumatic fever, lupus erythematosus, myocardial
infarction, lymphoma, leukemia, carcinoma, various liver diseases, and
many infections.[325,453,473,474,519,520,545,549,550]

Deviations from normal ranges of serum copper concentrations have
been found to be useful indicators for monitoring patients with acute
leukemia and other lymphomas. Tessmer et al.[548,550] studied the highly
significant relationship between the serum copper level and the percentage of blast cells in the marrow, generally a useful guide for leukemia therapy.

In Hodgkin's disease, a concentration of serum copper higher than
150 μg/100 ml strongly suggests active disease, except in the presence
of pregnancy, estrogen administration, or chronic inflammation.[609]
Similar observations were made for reticulum cell sarcoma, lymphosarcoma, and multiple myeloma. Twenty-three of 24 patients with generalized
disease had increased serum copper concentrations, whereas abnormally
high serum copper concentrations were less often seen when the process
was localized.[375] The concentration of serum copper in one patient with

multiple myeloma was an astounding 3,350 μg/100 ml, the highest ever reported, and the administration of penicillamine did not significantly increase urinary copper. The mechanism of this hypercupremia remained obscure.[203]

The differentiation of several types of liver disease sometimes can be achieved through the diagnostic use of radiocopper.[523]

Copper in Neoplasms

The copper content of benign tumors has been shown to be lower than that of carcinomas of the esophagus, bronchus, intestinal tract, and breast.[191] Sandberg noted large accumulations of copper and iron in the liver and spleen of patients with cancer of the respiratory system, genitourinary tract, or breast.[464] Pedrero reported that metastatic tumors in the liver, reticulum cell sarcoma not accompanied by hepatic disease, and diabetes mellitus were often associated with a low hepatic copper content,[418] but hepatic copper was found to be normal in patients with bronchogenic carcinoma.[372]

A possible relation between environmental zinc and copper and the occurrence of neoplasms was considered by Stocks and Davies, who examined these metals in garden soils in Wales, Cheshire, and Devonshire. They found the zinc:copper ratio consistently higher in gardens of persons who had died of cancer of the stomach.[528] In contrast, the experimental administration of 0.5% cupric oxyacetate in the diet of albino rats given maize with 0.09% 4-dimethylaminoazobenzene for 7 mo or longer afforded striking protection against the development of tumors.[160, 247]

Antimicrobial Effects

The effects of the combination of copper ions with proteins probably account for the low-grade antimicrobial activity of cupric sulfate (bluestone). A solution of about 0.05% has been used as a retention enema to treat typhoid fever and amebiasis. Cupric sulfate also has been applied topically for treatment of trachoma.[408] Metallic copper displays a gonococcicidal effect *in vitro* that may provide a degree of prophylaxis for women employing a copper-containing intrauterine device.[169] Copper sieves that form traps in an inhalation therapy circuit partially sterilize the vapor and reduce the incidence of pulmonary infection.[128]

Phosphorus Burns

Copper sulfate, applied in solution to the skin, is used to treat phosphorus burns because the formation of copper phosphide renders the phosphorus innocuous. If copper sulfate concentrations of 3% or more are used, hemolysis and death may occur.[243,534]

Emesis

Oral doses of 100–300 mg copper sulfate in water bring about especially effective emesis after the ingestion of phosphorus.[245,408] However, cupric sulfate is not a safe emetic. If vomiting does not occur after its use, gastrointestinal irritation, hemolysis, or other effects of acute toxicosis may supervene. Recently a patient given 2 g cupric sulfate as an emetic died from copper poisoning.[510]

Other Conditions

There is no evidence that copper metal or salts[570] are of value in the treatment of arthritis or epidermophytosis.[408]

SITUATIONS OF POTENTIAL COPPER TOXICOSIS

Copper-Containing Intrauterine Contraceptive Devices

Winding several hundred square millimeters of copper wire around a plastic intrauterine device (IUD) improves its contraceptive efficiency from 18.3 pregnancies to less than 1.0 pregnancy/100 woman-years of experience.[106,219,315,546,637] Analysis of such IUDs that have been *in utero* from months to years shows that about 25–30 mg copper is lost per year. Some of the metal is excreted with endometrial secretions, but studies in rats suggest that as much as 10–20 mg may be absorbed.[399] There is at least a possibility that such retention could lead to chronic toxicosis over the years or decades a woman is likely to use an IUD. The amount of copper absorbed from the uterus is comparable to that retained from dietary copper by the tissues of patients with Wilson's disease. Although in both cases the amount in question may only be a small fraction of dietary copper usually ingested, the parenterally absorbed copper from the IUD may not be metabolized and excreted by the same homeostatic mechanisms operating on orally ingested copper. Unfortunately, neither periodic determinations of blood or urinary copper nor clinical liver function tests can indicate whether copper is accumulating in the liver.

Therefore, quantitative analyses for copper as well as light and electron microscopic examinations of hepatic biopsy tissues from women who have used these devices for varying durations would be necessary to determine whether systemic toxicosis is occurring.

Copper-Supplemented Animal Feeds

Porcine liver is a principal constituent of some prepared meats, and much is eaten fresh. Pigs fed rations containing 250 ppm copper to accelerate growth (cf. Chapter 4) increase their hepatic copper from a normal mean of 24 μg to a mean of 220 μg/g dry tissue.[54] One-quarter lb (112 g) of liver from swine fed such diets may contain 10 mg of copper—an amount capable of causing acute toxicosis—or 2-3 times the average daily supply of the metal in a Western diet. Liver proteins bind at least a portion of the metal, mitigating the acute toxicosis, but no data on the effects of eating this amount of copper for long periods are available.

Manure from pigs raised on copper-supplemented feeds also constitutes a potential problem if it is used to fertilize land on which human food crops are grown: The copper content of vegetation and water runoffs from dressed fields may be increased.

Because of its antibiotic effect (see Chapter 4), copper also has been added to poultry feeds. Few data are available on the concentrations of tissue copper in chickens and turkeys.

Because of the possible ill effects on the environment and humans of adding 250 ppm copper to pig and poultry feed, the Food and Drug Administration currently limits the amount of copper for finished feeds to no more than 15 ppm.[386, 575]

6

Copper As an Industrial Health Hazard

OCCUPATIONAL EXPOSURE

The paucity of literature on ill effects caused by exposure to copper and its compounds in industry suggests that copper is not a particularly hazardous industrial substance. However, if workers are exposed to excess concentrations of the metal in any of its forms, undesirable health effects can result.[97] Copper melts and boils at high temperatures and does not give off metal fumes as readily as do more volatile metals such as lead, cadmium, and zinc. Dusts and fumes from copper and its compounds usually have an objectionable taste—a warning that tends to limit exposures before serious toxic intake can occur. However, metal fume fever from exposure to copper can occur.

Typical metal fume fever,[320,348] a 24–48-h illness characterized by chills, fever, aching muscles, dryness in the mouth and throat, and headache, was found in a copper refinery worker riveting heavy copper bus bars using a shielded-arc welding technique (personal communication, K. W. Nelson). Another worker contracted metal fume fever when he welded a copper tank.[170] Copper fever has been discovered among men handling copper oxide powder in a paint factory,[478] and copper acetate dusts have caused complaints of sneezing, coughing, digestive disorders, and fever.[60] Laborers handling "jewelry sweeps," a dusty scrap from jewelry manufacturing, experienced a bitter taste and nasal irritation traced to the verdigris formed from copper in the jewelry alloys (personal communication, K. W. Nelson). Apparent metal fume fever has also been reported in three men who were exposed to dust produced during the polishing of copper plates.[195]

Contact dermatitis associated with copper has been reported,[463] but few cases of dermatitis caused by copper metal or compounds occur in industry. Neither Stokinger[528a] nor Browning[68] in their comprehensive reviews of industrial toxicology mentions skin complaints, other than a green coloration noted more than a century ago among copper workers. A similar localized coloration is caused today from wearing jewelry made of copper or high copper alloys.

Observations of scores of copper smelter and refinery workers over the last 25 years have not revealed any significant incidence of dermatitis that could be traced to exposure to either copper or many of its inorganic compounds (personal communication, S. S. Pinto). Nor have chronic systemic effects from copper exposure been significant. The existence of such effects has been a subject of speculation, but no solid supporting evidence has been advanced.

In a recent review of health hazards from copper exposure, Cohen[97] observed that copper was ordinarily a benign agent. The combination of conditions in industry that would produce excessive concentrations of copper as a dust, fume, or mist, or in particle sizes and chemical forms such that toxic effects would be generated from the copper absorbed, is relatively rare.

The U.S. Occupational Safety and Health Administration has adopted standards for exposure to airborne copper at work. The time-weighted average for 8-h daily exposures to copper dust is limited to 1.0 mg/m³ air. The standard for copper fume was changed in 1975 to 0.2 mg/m³. No particle size or solubility specifications are included in the standards, which were derived from threshold limit values (TLV) adopted by the American Conference of Governmental Industrial Hygienists. Documentation for the TLVs consists only of Gleason's research[195] and a personal communication.

Four studies have found increased incidences of lung cancer among workers in copper smelters.[300,305,361,565] The authors have suggested that the cancer was caused by exposure to arsenic trioxide in dust and fumes produced by the various pyrometallurgic processes. They did not suggest that copper itself played any etiologic role in the cancer deaths.

COMMUNITY EXPOSURES

Water

A 1969 Public Health Service study of 969 urban water supply systems revealed that 11 supplies contained copper in concentrations above the drinking water standard of 1 ppm, a standard based on taste.[580] The maximum concentration found was 8.35 ppm. Copper in public water supplies has not been treated by regulatory agencies as a significant problem. Indeed, copper is intentionally added to the New York City water supply to maintain a concentration of 0.059 ppm, which controls algal growths.[285]

Air

The National Air Sampling Network's (NASN) 1966 data indicate a range of airborne copper concentrations from 0.01 to 0.257 $\mu g/m^3$ in rural and urban communities.[581] Continuous monitoring of air near copper smelters for over 10 years usually has shown fractional $\mu g/m^3$ concentrations. Occasionally weekly averages of 1-2 μg are reported (personal communication, K. W. Nelson). Even when airborne copper does reach this level, the dose of the metal would be about 1% of the normal daily ingested dose, given a 15 m^3 daily intake of air and a total penetration, retention, and absorption of all airborne copper. Schroeder reached the same quantitative conclusion in reviewing NASN data.[479]

It should be noted that the validity of all airborne copper measurements derived from samples collected with conventional high-volume sampling equipment has been questioned.[238] Copper abraded from motor commutators may have contaminated the air around the sampling units.

Most copper emissions in the United States are produced by the metallurgic processing of ores and concentrates.[122] Sources of copper-bearing dust and fume in smelters are roasters, reverberatory furnaces, and converters. The typical particulate collection systems are made up of large balloon flues for gravity separation of the coarser dusts and fume agglomerates, and electrostatic precipitators with collection efficiencies of 95-99%.

The second most important source of copper emissions is the iron and steel industry.[122] Trace quantities of copper enter the steelmaking process in raw materials. Emissions are generated mostly from blast furnaces and open hearth furnaces; the emitted dusts and fumes contain 0.1-0.5% copper. Controls are a combination of cyclone separators and electrostatic precipitators.

Power plants that burn coal are the third most important source of copper emissions. Based on measurements by Cuffe, the average concentration of copper in particulates in power plant stack gases (with electrostatic precipitators) is 230 μg/m^3.[107] Emissions from plants without emission controls would be about 7 times higher.

Other noteworthy emission sources are brass and bronze foundries, secondary smelting of copper and its alloys, burning of insulation from copper wire, and miscellaneous fabricating operations. Because of the low magnitude of emissions from these sources and the minimal environmental impact of copper, not much general data have been accumulated on such emissions.

Copper Emission and Ambient Air Standards

Because the economic value of copper encourages its capture from process gases, general air pollution controls are used to prevent significant mass emissions of copper. Atmospheric levels of copper have not been proven to pose a risk to human health; hence, no emission or ambient air standards for copper have been established or proposed.

7

Summary and Conclusions

SUMMARY

Copper in the Ecosystem

Almost 2 million metric tons of copper are removed from the sites of
their natural sources and injected into the world ecosystem annually.
The concentration of copper in the continental crust is about 50 ppm.
Most soils, plants, and many surface and ground waters contain 1 or
more ppm copper. The total body content of copper in adult mammals
is about 2 ppm wet weight.

Supplementing the copper in an animal's feed with appropriate levels
of iron and zinc has been thought to increase growth rates. Because
such supplements also will produce manure containing as much as
8,000 ppm copper—potentially harmful to soils, crops, animals, and
humans—care is required in their disposal.

Copper in Plants

Several specific copper proteins have been isolated from plant tissues
and characterized chemically.

Copper is essential to the normal growth and development of almost all plants. Plants grown in soils that contain less than about 5 ppm are likely to show adverse effects. Copper toxicosis in plants rarely is observed under natural conditions but may occur where large amounts of copper have been added to the soil. The absolute concentrations of copper that result in pathologic deficiency or excess depend upon the species of plant and the physicochemical characteristics of the soil. For a number of food species, supplementation of soil, seeds, or the whole plant with copper can enhance crop yields.

Copper in Animals

Copper deficiency can be produced experimentally in many animal species, but naturally occurring, clinically significant deficiency generally is limited to cattle and sheep. Cattle are more susceptible than sheep, and monogastric animals rarely are subject to copper deficiency. Copper toxicosis also can be induced in many species, but naturally occurring toxicosis, like deficiency, commonly occurs only in sheep and cattle. Sheep are more susceptible than cattle to copper toxicosis. Again, nonruminant animals are much more resistant to copper toxicosis. It should be understood that the amounts of copper required to prevent deficiency or cause toxicosis in animals may vary appreciably with the amounts of zinc, iron, molybdenum, and sulfate in the diet.

The differences between cattle and sheep make it highly advisable that feeds and mineral supplements be differently formulated for each species. It has been reported that supplementing feeds with high levels of copper may quicken the rate of weight gain in young pigs and chickens, but the evidence is inconclusive.

Some aquatic organisms, including edible fish, are susceptible to toxicosis by copper concentrations two orders of magnitude lower than the accepted standard for drinking water (1.0 ppm). Copper is an effective molluscicide and is useful in the control of schistosomiasis.

Human Copper Metabolism

Copper is essential to normal health and longevity in man. It is the prosthetic element of more than a dozen specific copper proteins.

In relation to strict metabolic requirements, copper is overabundant in almost all human diets. Therefore, clinically significant copper deficiency is extremely unusual and is virtually limited to instances of severe gastrointestinal malabsorption, drastically reduced dietary intake (and even this condition is significant only in newborn infants), or to the presence of a rare X-linked disorder of copper absorption and transport known as Menkes's steely- or kinky-hair disease. Human copper toxicosis is also extremely rare, and appears in clinically significant form almost only when suicide is attempted by the ingestion of large quantities of a copper salt, or where a genetic defect in copper metabolism is inherited in an autosomal recessive fashion (Wilson's disease). In patients with Wilson's disease, copper steadily accumulates, first in the liver and then in other parts of the body. Damage, particularly evident in the liver and central nervous system, is ultimately fatal. Successful treatment and prophylaxis is effected by a chelating agent, D-penicillamine, which promotes the urinary excretion of copper.

Genetic mechanisms are clearly efficient in regulating the balance of dietary copper, but there is little knowledge of whether they act on parenterally introduced copper. Copper introduced during hemodialysis or into the uterus as a contraceptive is absorbed systemically to some degree. There is evidence that copper also may be absorbed parenterally through the skin, lungs, and uterine mucosa. It is not known whether this copper accumulates or is excreted. Hepatic and pulmonary granulomas and neoplasms have been observed in vineyard workers exposed to sprays of copper sulfate solutions.

Alterations in copper metabolism, reflected in the concentrations of copper and ceruloplasmin in the serum, are associated with pregnancy and the administration of estrogens. Changes in these concentrations also accompany many acute and chronic disorders.

Copper in Drinking Water

Deficient and excessive copper in soil and water will greatly affect agriculture, animal husbandry, and the economic and biologic aspects of ppm, unless water solutions of low pH are allowed to stand for a long time in such plumbing.

Copper As an Industrial Health Hazard

Although copper can act as a toxic agent in an occupational setting, it is benign under ordinary circumstances. However, if workers are exposed to excessive concentrations of the metal in any of its forms, there may be undesirable health effects. Because of the absence of reports on significant environmental effects from airborne copper, copper and its compounds as dusts or fumes dispersed into the atmosphere have not been considered hazardous.

CONCLUSIONS

Copper in the Ecosystem

Copper should be used with awareness of its ultimate distribution and effects on the ecosystem.

Copper in Plants

Judicious use of copper may aid in obtaining optimal yields of crops and inhibiting the growth of undesirable plants, particularly fungi.

Copper in Animals

Deficient and excessive copper in soil and water will greatly affect agriculture, animal husbandry, and the economic and biologic aspects of certain aquatic organisms.

Human Copper Metabolism

Copper is essential to the life and health of human beings, and the interaction of its environmental supply with the mechanisms controlling its absorption, transport, and excretion is so finely tuned that significant clinical manifestations of deficiency or toxicosis are very rare.

8

Recommendations

1. Research into the mechanisms of interaction between copper and molybdenum, sulfate, iron, and zinc in plant and animal metabolism is desirable.

2. The optimal dietary requirements of copper, molybdenum, sulfate, iron, and zinc for the various species of animals that are sources of human food should be determined.

3. A system for verifying and tabulating incidents of deficiency and excess of copper and interrelated trace elements in animals should be initiated on a national basis.

4. Copper should not be generally recognized as safe for livestock feeds without qualification.

5. Copper should continue to be added to livestock and poultry feeds only in the concentration (15 ppm) generally regarded as safe. However, because of widespread use of high level (250 ppm) copper supplementation in animal feeds in the United States and elsewhere, the beneficial and harmful effects of such supplementation should be further investigated. This inquiry should include a careful monitoring of the disposal of animal wastes.

6. Research directed at understanding better the biochemistry and physiology of copper proteins should be encouraged and supported.

7. Because copper is absorbed by the lungs, skin, and uterus, as well as by the gastrointestinal tract, a nationwide clinical investigation should be carried out to determine whether any long-term hazard of human copper toxicosis is possible from the added burden of body copper introduced parenterally through chronic hemodialysis inhalation, or absorption from the skin, or copper-containing intrauterine contraceptive devices.

8. The role, if any, of copper in producing granulomas or malignant tumors, particularly in liver and lungs, should be defined.

9. Studies defining the role of copper as an etiologic agent in metal fume fever should be carried out.

Appendix A

Copper Analysis in Environmental and Biologic Samples

Selected methods for the quantitative analysis of copper in the environment (water and air), in biologic materials, and animal feeds, and of ceruloplasmin in human serum are described in this appendix. No attempt has been made to note and compare all the analytic procedures available. Instead, methods known to be feasible and accurate by the members of the Subcommittee are presented.

Techniques most commonly used to analyze environmental samples for copper are noted. The analysis of water is emphasized, because airborne particulate (as well as other environmental samples) may often be assayed by analysis of aqueous solutions or suspensions.

A lengthy description of atomic absorption is presented because this method is accepted as standard for copper by the American Society for Testing and Materials (ASTM), the Environmental Protection Agency, and other organizations that have presented standard methods. The method has a sensitivity of .04 μg/l as presented, but the optimal concentration range may be varied by changing the analytic line used to suit a particular sample. Most atomic absorption measurements have an accuracy of approximately 2%.

WATER

Atomic absorption spectroscopy is similar to flame emission photometry in that a sample is atomized and aspirated into a flame. Flame photometry, however, measures the amount of light emitted, whereas in atomic absorption spectrophotometry, a light beam is directed through the flame into a monochromator, and then onto a detector that measures the amount of light absorbed. In many instances, absorption is more sensitive because it depends upon the presence of free unexcited atoms, and even at flame temperatures the ratio of unexcited to excited atoms is very high. Since the wavelength of the light passed by the monochromator is selected to be characteristic of the difference between two energy levels of the metal being determined, the light energy absorbed by the flame is a measure of the concentration of that metal in the sample. This principle forms the basis of atomic absorption spectroscopy.

In determining copper concentrations, contamination and loss are of prime concern. Dust in the laboratory environment, impurities in reagents, and impurities on laboratory apparatus that the sample touches are all sources of potential contamination. For liquid samples, containers can introduce either positive or negative errors in the measurement of trace metals by contributing contaminants through leaching or surface desorption and by depleting them through adsorption. Thus the collection and treatment of the sample prior to analysis require particular attention. The sample bottle should be thoroughly washed with detergent and tap water; next it should be successively rinsed with 10% hydrochloric or nitric acid, and three times with distilled or demineralized water. Before collecting the sample, it should be decided what type of data is desirable; that is, dissolved, suspended, total, or extractable.

For the determination of dissolved copper, the sample should be filtered through a 0.45-μm membrane filter as soon as practical after collection. Use the first 50–100 ml of filtrate to rinse the filter flask. Discard this portion and collect the required amount of filtrate. Acidify the filtrate with 1:1 redistilled nitric acid (3 ml/l). Normally, this amount of acid will lower the pH to 2 or 3 and should be sufficient to preserve the sample indefinitely. Analyses performed on a sample so treated should be reported as "dissolved" concentrations.

To determine suspended copper, a representative volume of sample should be filtered through a 0.45-μm membrane filter. When considerable sediment is present, as little as 100 ml of a well-shaken sample is filtered. Record the volume filtered and transfer the membrane filter containing the sediment to a 250-ml Griffin beaker and add 3 ml distilled nitric acid. Cover the beaker with a watch glass and heat gently. The warm acid will soon dissolve the membrane. Increase the temperature of the hotplate and digest the material. When the acid has evaporated, cool the beaker and watch glass and add another 3 ml distilled nitric acid. Cover and continue heating until the digestion is complete, generally indicated by a light-colored residue. Add 2 ml distilled 1:1 hydrochloric acid to the dry residue and warm the beaker again gently to dissolve the material. Wash down the watch glass and beaker walls with distilled water and filter the sample to remove silicates and other insoluble material that could clog the atomizer. Adjust the volume to some predetermined value based on the expected concentrations of the metal present. This volume will vary according to the metal being determined. The sample is now ready for analysis. Concentrations so determined should be reported as "suspended." Quantities of copper determined on unused membrane filters should be deducted from the total quantity found. Ordinarily such amounts are insignificant.

To determine total copper, the sample is not filtered before processing. Choose an amount of sample appropriate for the expected level of the metal. If much suspended material is present, as little as 50–100 ml of well-mixed sample will most probably be sufficient. (The sample volume required may vary proportionally with the number of metals to be determined.)

Transfer a representative aliquot of the well-mixed sample to a Griffin beaker and add 3 ml concentrated distilled nitric acid. Place the beaker on a hotplate and evaporate to dryness, making certain that the sample does not boil. Cool the beaker and add another 3 ml portion of distilled concentrated nitric acid. Cover the beaker with a watch glass and return to the hotplate. Increase the temperature of the hotplate so that a gentle reflux action occurs. Continue heating, adding additional acid as necessary until the digestion is complete, generally indicated by a light-colored residue. Add sufficient distilled 1:1 hydrochloric acid and warm the beaker again to dissolve the residue. Wash down the beaker walls and watch glass with distilled water and filter the sample to remove silicates and other insoluble material that could clog the atomizer. Adjust the volume to some predetermined value based on the expected metal concentrations. The sample is now ready for analysis. Concentrations so determined should be reported as "total."

Optimal Concentration Range   0.1–10 mg/1
Wavelength   324.7 nm
Sensitivity   0.04 mg/1
Detection Limit   0.005 mg/1
Preparation of Standard Solution

1. Stock solution: Carefully weigh 1.0 g electrolytic copper (analytic reagent grade). Dissolve in 5 ml redistilled nitric acid and make up to 1 liter with distilled water. Final concentration is 1 mg copper/ml (1,000 mg/l).

2. Prepare dilutions of the stock solution to be used as calibration standards at the time of analysis. Maintain an acid strength of 0.15% nitric acid in all calibration standards.

General Instrument Requirements[582]
1. Copper hollow cathode lamp
2. Wavelength: 324.7 nm
3. Type of burner: Boling
4. Fuel: acetylene
5. Oxidant: air
6. Type of flame: oxidizing
7. Photomultiplier tube: IP-28

Herrman and Lang[233] first described atomic absorption analysis for copper in 1963, and Berman[34] has reviewed copper analysis by atomic absorption in tissues and biologic samples. Dispersion of copper in methyl isobutyl ketone will increase the test sensitivity about four times. The high temperature at which copper is volatilized (600°C) permits ashing of samples, an action that will remove several interfering substances adequately. Tissue, feed, grain, forage, and other materials have been analyzed easily using atomic absorption spectrophotometry.[34,505]

Other techniques may be employed for determining copper in aqueous solution or suspension, including spectrophotometry (applied below to biologic samples), polarography, and anodic stripping voltammetry.

AIR

Samples of airborne particulates preferably are collected on glass fiber or membrane filters. Ideally, particle size distributions and identifications or chemical compounds should be attempted, but this is impractical for any routine monitoring. However, a rough separation of respirable and irrespirable particles is practical and provides useful information.

The analytic method of choice is atomic absorption after acid digestion of the filter and appropriate dilution. Spectrographic, polarographic, spectrophotometric, neutron activation analysis, and anodic stripping voltammetry may also be used. High sensitivity, accuracy, and precision are easily attainable.

Microanalysis of Biologic Materials

This procedure, a combination and modification of two procedures described in the literature,[148,371,426] is used to determine quantitatively the total copper content of organic material by using dicyclohexanone-oxalyldihydrazone (DCO).

Reagents
1. Concentrated sulfuric acid, reagent grade of the American Chemical Society (ACS).
2. Perchloric acid, 60%, reagent ACS.
3. Ammonium hydroxide, reagent ACS.
4. Phosphate-citrate buffer, pH 7.4, made of 900 ml 0.4 M anhydrous dibasic sodium phosphate, and 100 ml 0.2 M monohydrate citric acid.
5. DCO reagent, prepared by dissolving 0.1% DCO (Eastman Organic 7175)* in hot 50% (vol/vol) ethanol. *Do not heat reagent and ethanol together.*
6. Standard copper solutions containing 1.0 and 2.0 μg copper/ml are made in 0.10 M sulfuric acid.

Equipment
1. Spectrophotometer: Zeiss PMQ II, Beckman DU, or equivalent instrument with attachments for use of cuvettes of 40 or 50 mm path length.
2. Digestion apparatus: Microdigestion shelf, gas heated, 6-unit, with Pyrex glass fume duct (Fisher Scientific 21-130).
3. Digestion tubes: Tubes, similar in form to blood sugar tubes, are made by Robert C. Ewald, Inc., Middle Village, N.Y. Volume of the bulb, approximately 6.5 ml; constricted part of the tube graduated at 7.0 ml (just above the bulb), 8.0 ml, and 9.0 ml; total height of the tube, 21 cm; diameter of the upper portion of the tube, 2 cm; diameter and length of the constricted portion of the tube, 1 cm and 5.5 cm, respectively.

All glassware must be washed free of copper with 10% hydrochloric acid and rinsed with a large volume of distilled or de-ionized water.

Procedure Blanks, standards, and unknown samples of biologic materials are run in triplicate where possible. Pipet the following solutions into digestion tubes that contain three glass beads.

*Specific products have been listed solely to help readers who desire more information. Mention of these products does not constitute an endorsement by the National Academy of Sciences or the National Research Council.

1. Blank: 1.0 ml of 0.10 M sulfuric acid.
2. Standards: 1.0 ml of 1.0 and 2.0 $\mu g/ml$ copper solutions.
3. Unknown samples:
   - Serum or plasma: 1.0 ml.
   - Urine: 1.0–3.0 ml, depending on expected concentration of copper. For urine specimens of patients not on chelation therapy, 3.0 ml is appropriate.
   - Tissue: For this assay, 50–100 mg of dried tissue is usually sufficient. If much less material is available, the procedure can be modified (see below). Add 1.0 ml of de-ionized water. Add 1.0 ml of concentrated sulfuric acid to all tubes and mix. Add 1.0 ml of 60% perchloric acid and mix.

Tubes are heated in the digestion apparatus and shaken gently until the onset of boiling. Particles of carbon ascend to about the middle of the tube, eventually to be washed down by the refluxing liquid. The digest turns colorless, then yellow, then colorless again. At this point, sulfuric acid starts to reflux, and should be continued for 15 min. Total time of digestion is approximately 30 min.

After cooling, 1.0 ml distilled water is added and the solution is mixed and cooled again. From a burette, 3.5 ml concentrated ammonium hydroxide is added slowly while the tube is cooled in ice water. Tubes are placed in a water bath maintained at 65–70°C for 17–18 h. Under these conditions practically all the free ammonia, but no ammonium ion, is removed from the solution.

To each tube, add 2.0 ml phosphate-citrate buffer and dilute with water to the 8.0-ml mark. This addition brings the contents of the tube within the pH limits of 7–8 necessary for color to develop.

Add 0.8 ml DCO reagent to each tube, cover with Parafilm, and mix thoroughly.

After 1 h at room temperature, the optical density of each solution is read against water at 600 nm in the spectrophotometer using 40 or 50 mm light-path cuvettes.

If the optical density of a solution is greater than 0.600, the sample should be diluted with phosphate-citrate buffer and more DCO reagent, equal to 10% of the volume of sample and buffer, should then be added. Alternatively, the unknown sample should be redigested, using a smaller amount.

Calculations and sources of error   The copper content of each unknown is calculated by comparing its net optical density to the net optical density of the standard.

The standard deviation (sd) of analyses performed by this method over the range of 1.4–3.7 μg copper/ml was estimated to be 0.0392 μg copper/ml, according to data from seven sets of triplicate measurements.

There are three common sources of error in this procedure: specimens have been contaminated with copper; copper may be lost if bumping occurs during digestion; or the pH of the final solution may be outside the 7–8 range.

Modifications for Applications of Method for Needle Biopsy Specimens of Liver

1. Standards: 1.0 ml of 0.25 μg/ml and 1.0 μg/ml of copper.
2. Digestion: Add to blank, standard, and unknown tubes:
   0.3 ml concentrated sulfuric acid
   0.3 ml 60% perchloric acid
   Digest until sulfuric acid refluxes for 10 min.
3. Neutralization: To the cooled digested sample, add 0.5 ml distilled water and 1.1 ml concentrated ammonium hydroxide. Place in 90°C water bath until odor of the ammonium hydroxide disappears (about 1½–2 h). Wash the contents of the digestion tube into a test tube calibrated at 2.7 ml, using a total of 1.0 ml of phosphate-citrate buffer that has been diluted 1:1 with distilled water. Add water to the 2.7 mark.
4. Development of color: Add 0.3 ml dco reagent and mix. Read optical density after 1 h.

Range of Values

Serum or plasma The plasma of Americans and Europeans contains approximately 100 μg copper/100 ml,[473] most of which is tightly bound to ceruloplasmin. The range of copper concentration in the plasma of normal men is 81–137, and in normal women, 87–153 μg/100 ml.[87]

Concentrations of serum copper below the lower limit of the normal range are found in patients with hereditary Wilson's or Menkes's diseases, as well as in five acquired pathologic conditions: the nephrotic syndrome, kwashiorkor, sprue, scleroderma of the intestine, and protein-losing enteropathy. Physiologic hypocupremia is present during the first four to six months of life in almost all infants.

Concentrations of serum copper above the upper limit of normal ranges are common in late pregnancy and following the ingestion of estrogens or contraceptive pills; high concentrations are also found in many inflammatory, necrotizing, and neoplastic diseases.[473]

Urine   The normal 24-h excretion of copper in the urine is less than 30 μg. Patients with Wilson's disease who have received no treatment usually excrete considerably more than 100 μg/24 h.[473] Treatment with appropriate pharmacologic agents will greatly increase the amount of copper excreted.

Hepatic copper   The normal mean ± SD concentration of hepatic copper is 31.5 ± 6.8 μg/g dry liver. Hepatic copper concentrations of untreated patients with Wilson's disease measure more than 250 μg/g dry liver.

Measurement of the Concentration of Ceruloplasmin in Human Serum

Ceruloplasmin is a copper-containing globulin of plasma that can catalyze the oxidation of *p*-phenylenediamine (PPD).[240,242,370] The rate at which PPD is oxidized is proportional to the concentration of ceruloplasmin in serum. When determined under precisely defined conditions of composition of the medium, and at a given temperature, the rate of oxidation allows the calculation of the concentration of enzyme in the serum. In the method described below, the rate of PPD oxidation is measured by quantitatively determining the rate of darkening of its solution in a spectrophotometer.

The major diagnostic value in measuring the concentration of serum ceruloplasmin is in suspecting, confirming, or ruling out the diagnosis of Wilson's disease.[476,520]

As described here, this method has been calibrated only for human serum.

Reagents
1. Acetate buffer. Dissolve 10.05 g sodium chloride and 49.20 g anhydrous sodium acetate in about 1,900 ml distilled water. Adjust pH to 5.12 with about 10 ml glacial acetic acid. Bring volume to 2,000 ml with distilled water. Buffer is stable indefinitely at 4°C.
2. PPD reagent. A 0.5% solution of *p*-phenylenediamine dihydrochloride in acetate buffer that has been warmed to 30°C is prepared immediately before its addition to the cuvette.

Equipment
1. Zeiss Spectrophotometer PMQ II.
2. Water bath circulator (Bronwill Circulator, Will Corp., N.Y.)
3. Electronic therometer with flexible probe (Tri-R Instruments).
4. Electric timer or stopwatch.

5. Hotplate.
6. Cuvettes, 1 cm path length.

Procedure   Water is circulated through the cell compartment of the spectrophotometer and the temperature is adjusted so that the reading taken in a reference cuvette containing water is 30 ± 0.1°C. The enzymatic activity of ceruloplasmin in this system is increased or decreased respectively by about 1% for each 0.1°C rise or fall in temperature.

1. Place 1.0 ml distilled water in reference cuvette and 1.0 ml fasting, nonhemolyzed serum in the sample cuvette.

2. Dissolve weighed PPD in an appropriate volume of warm buffer, and add 2.0 ml of this warm PPD reagent to the serum in the sample cuvette.

3. Add 2.0 ml of warmed buffer to the reference cuvette.

4. Cover cuvettes with Parafilm and mix by inverting them.

5. Warm unknown sample quickly to 30°C on hotplate, and begin taking readings about 3 min after mixing.

At 530 nm, readings of optical density of the unknown are made against the reference cuvette at intervals of 30 sec–2 min, depending on the number of samples and the concentration of ceruloplasmin. A sufficient number of readings is made so that the last 5 or 6 points fall on a straight line. When plotted against time, that is, the change in optical density/min/ml, the slope of this line is proportional to the ceruloplasmin content of the serum. Be sure to check the temperature of sample at end of run.

Calculations and sources of error   If x is the ceruloplasmin concentration in mg/100 ml of serum, and y is ΔOD^*/min/ml − 0.0012, then $x = 900y$.

This method has been calibrated with human serum of known ceruloplasmin copper content; the copper content of ceruloplasmin was assumed to be 0.31%.[269] Serum copper was determined by wet digestion of a sample of serum from which nonceruloplasmin copper (approximately 5% of total serum copper) was removed by addition of sodium diethyldithiocarbamate and passage through a column of activated charcoal.[518] Hemolyzed, icteric, lipemic, aged, or frozen and thawed specimens may yield unsatisfactory assays. Changes in temperature during assays will also produce errors.

Range of values   In a series of 185 unselected normal adult subjects,

---

*Change in optical density.

the mean ± SD ceruloplasmin concentration was 30.5 ± 3.5 mg/100 ml of serum.

Almost all patients with Wilson's disease exhibit ceruloplasmin concentrations of 0-20 mg/100 ml of serum. Decreased serum concentrations also have been found in about 20% of healthy heterozygous carriers of 1 abnormal "Wilson's disease gene"[476] and in all newborns during the first 6 months of life.

Pathologic deficiency of serum ceruloplasmin may occur in the nephrotic syndrome, kwashiorkor, sprue, scleroderma of the intestine, protein-losing enteropathy, Menkes's disease, and in rare instances of severe hepatitis.

Increased concentrations of serum ceruloplasmin have little diagnostic significance, since they are encountered late in pregnancy and following the ingestion of estrogens or contraceptive pills, and in many inflammatory, necrotizing, and neoplastic diseases.[492,520]

Appendix B

Glossary of Compounds

Name	Formula
Acetic acid	$CH_3 \cdot COOH$
Adenosine triphosphate (ATP)	$C_{10}H_{12}N_5O_3H_4P_3O_9$
Ammonia	NH_3
Ammonium hydroxide	NH_4O_4
Ammonium molybdate	$(NH_4)_2MoO_4$
Arsenic trioxide	AS_2O_3
L-Ascorbic acid	$C_6H_8O_6$
p-Aminosalicylic acid	$C_7H_7NO_3$
Aureomycin (chlortetracycline hydrochloride)	$C_{22}H_{23}ClN_2O_8 \cdot HCl$
British antilewisite (BAL)	$CH_2SHCHSHCH_2OH$
Calcium carbonate	$CaCO_3$
Catechol	$C_6H_4(OH)_2$
Citric acid, monohydrate	$C_6H_8O_7 \cdot 7H_2O$
Copper (II) acetylacetonate also known as *bis* (2,4-pentanedionato) copper	$[(H_3CCO)_2CH]_2Cu$
Copper dimethyldithiocarbonate also known as *bis* (dimethyldithiocarbamato) copper	$[(CH_3)_2NC(S)_2]_2Cu$
Copper orthophosphate	$Cu_3(PO_4)_2 \cdot 3H_2O$
Copper pentachlorophenate also known as *bis* (pentachlorophenolato) copper	$(C_6HCl_5O)_2Cu$
Copper 3-phenylsalicylate also known as 3-phenylsalicylic acid, copper (II) salt	$[C_6H_5(C_6H_3OHCOOH)]_2 \cdot Cu$
Copper phosphide	Cu_3P
Copper pyrophosphate also known as pyrophosphoric acid, copper (II) salt	$H_4P_2O_7 \cdot 2Cu$

75

Name	Formula
Copper resinate	Of indeterminate composition
Copper ricinoleate	$[(CH_2)_5CH(OH)CH_2CH=$
also known as ricinoleic acid, copper (II) salt	$CH(CH_2)_7COOH]_2 \cdot Cu$
Cupric acetate, hydrate	$Cu(CH_3COO)_2 \cdot H_2O$
Cupric acetoarsenite	$Cu(C_2H_3O_2)_2 \cdot 3Cu(AsO_2)_2$
Cupric carbonate, basic	$CuCO_3 \cdot C_4(OH)_2$
Cupric chloride	$CuCl_2$
Cupric gluconate, hydrate	$Cu[CH_2OH(CHOH)_4CO_2]_2 \cdot H_2O$
Cupric oxide	CuO
Cupric oxyacetate	$CuC_2O_3H_4$
Cupric sulfate	$CuSO_4$
Cupric tartrate	$CuC_4H_4O_6$
Cuprous chloride	$CuCl$
Cuprous iodide	CuI
Cuprous oxide	Cu_2O
Dehydroascorbic acid	$C_6H_6O_6$
Dicyclohexanoneoxalyldihydrazone (DCO)	$C_6H_{10}:NNHCOCONHN:C_6H_{10}$
4-Dimethylaminoazobenzene	$C_6H_5N_2C_6H_5N(CH_3)_2$
α-, α'-Dipyridyl	$C_{10}H_8N_2$
Dopamine	$C_8H_{11}NO_2$
Epinephrine	$(HO)_2C_6H_3CH(OH)CH_2NHCH_3$
Ethanol	C_2H_5OH
Ethylenediaminetetraacetic acid (EDTA)	$N_2(CH_2)_2(CH_2COOH)_4$
Galactose	$HOCH_2(CHOH)_4CHO$
Glucosamine	$HOCH_2(CHOH)_3CH(NH_2)CHO$
Glutathione	$HOOC \cdot CH(NH_2) \cdot (CH_2)_2 \cdot CO \cdot NH - CH(CH_2 \cdot SH) \cdot CO \cdot NH \cdot CH_2 \cdot COOH$
Heme	$C_{34}H_{32}N_4O_4Fe$
Hydrochloric acid	HCl
Hydrogen peroxide	H_2O_2
Hydroquinone	$C_6H_4(OH)_2$
Lysine	$H_2N(CH_2)_4CH(NH_2)COOH$
Mannose	$CH_2OH(CHOH)_4CHO$
Methyl isobutyl ketone	$CH_3COCH_2CH(CH_3)_2$
Nitric acid	HNO_3
Nitrilotriacetic acid (NTA)	$N(CH_2COOH)_3$
Norepinephrine	$(HO)_2C_6H_3CH(OH)CH_2NH_2$
D-Penicillamine	$(CH_3)_2C(SH)CH(NH_2)COOH$
p-Phenylenediamine	$C_6H_4(NH_2)_2$
p-Phenylenediamine dihydrochloride	$C_6H_4(NH_2)_2 \cdot 2HCl$
Perchloric acid	$HClO_4$
Protoporphyrin	$C_{32}H_{32}N_4(COOH)_2$
Serotonin	$HOC_8H_5N(CH_2)_2NH_2$
Sodium acetate, anhydrous	$C_2H_3NaO_2$
Sodium diethyldithiocarbamate	$(C_2H_9)_2N \cdot CS_2Na \cdot 3H_2O$
Sodium phosphate, dibasic	Na_2HPO_4
Sulfuric acid	H_2SO_4
Triethylene tetramine dihydrochloride	$C_6H_{18}N_4 \cdot 2HCl$
Tyrosine	$HOC_6H_4CH_2CH(NH_2)COOH$

References

1. Acute copper poisoning—Arizona. Morbid. Mortal. Week. Rep. 23:407, 1974.
2. Acute copper poisoning—Pennsylvania. Morbid. Mortal. Week. Rep. 24:99, 1975.
3. Adamson, A. H., and D. A. Valks. Copper toxicity in housed lambs. Vet. Rec. 85:368-369, 1969.
4. Aksenova, N. P. Effect of day length on oxidase activity in plants. Soviet Plant Physiol. (Translation Fiziol. Rasten.) 10:132-139, 1963.
5. Albiston, H. D., L. B. Bull, A. T. Dick, and J. C. Keast. A preliminary note on the aetiology of enzootic jaundice, toxaemic jaundice, or "yellows," of sheep in Australia. Austral. Vet. J. 16:233-243, 1940.
6. Aldinger, L. M. The effect of high copper levels on turkey performance. Poult. Sci. 45:1065-1066, 1966. (abstract)
7. Aliev, D. A. Effect of the combination of trace elements with nitrogen and phosphorus on the yield and quality of eggplant. Izv. Akad. Nauk Azerb. SSR, Ser. Biol. Nauk 1969(6):3-10. (in Russian)
8. Allcroft, R., K. N. Burns, and G. Lewis. Effect of high levels of copper in rations for pigs. Vet. Rec. 73:714-718, 1961.
9. Al-Rashid, R. A., and J. Spangler. Neonatal copper deficiency. New Engl. J. Med. 285:841-843, 1971.
10. Ammerman, C. B. Recent developments in cobalt and copper in ruminant nutrition: A review. J. Dairy Sci. 53:1097-1107, 1970.
11. Ammerman, C. B., F. G. Martin, and L. R. Arrington. Mineral contamination of feed samples by grinding. J. Dairy Sci. 53:1514-1515, 1970.
12. Arthur, J. W., and E. N. Leonard. Effects of copper on *Gammarus pseudolimnaeus, Physa integra,* and *Campeloma decisum* in soft water. J. Fish. Res. Bd. Can. 27: 1277-1283, 1970.

13. Ashkenazi, A., S. Levin, M. Djaldetti, E. Fishel, and D. Benvenisti. The syndrome of neonatal copper deficiency. Pediatrics 52:525-533, 1973.

14. Baker, D. E. Copper: Soil, water, plant relationships. Fed. Proc. 33:1188-1193, 1974.

15. Bal, H. Effects of copper sulphate poisoning in white rats. Naturwissenschaften 51:139, 1964.

16. Balla, F., M. Kiszel, and K. Gellért. Decomposition of vitamin C and the prevention thereof in food processing, with special respect to new methods of production. (Part IV) Élelmezési Ipar 14: 294-296, 1960. (in Russian)

17. Barber, R. S., R. Braude, and K. G. Mitchell. High copper mineral mixtures for fattening pigs. Chem. Ind. (London) 1955:601-602.

18. Barbour, B. H., M. Bischel, and D. E. Abrams. Copper accumulation in patients undergoing chronic hemodialysis. The role of Cuprophan. Nephron 8:455-462, 1971.

19. Barden, P. J., and A. Robertson. Experimental copper poisoning in sheep. Vet. Rec. 74:252-256, 1962.

20. Barka, T., P. J. Scheuer, F. Schaffner, and H. Popper. Structural changes of liver cells in copper intoxication. A.M.A. Arch. Path. 78:331-349, 1964.

21. Barranco, V. P. Eczematous dermatitis caused by internal exposure to copper. Arch. Derm. 106:386-387, 1972.

22. Barth, R., O. T. Goday, and G. Hauila. Observations on the nannoplankton and the concentration of copper in the Brazil current. Public. Inst. Pesq. Mar. 3:1-11, 1967. (in Portuguese)

23. Bauman, L. K. The copper content in tissues of patients with hepatolenticular degeneration. Zh. Nevropatol. Psikhiatr. 60:1141-1145, 1960. (in Russian)

24. Beard, A. W. The association of hepatolenticular degeneration with schizophrenia. Acta Psychiatr. Neurol. Scand. 34:411-428, 1959.

25. Bearn, A. G. Wilson's disease, pp. 1033-1050. In J. B. Stanbury, J. G. Wyngaarden, and D. S. Fredrickson, Eds. The Metabolic Basis of Inherited Disease. (3rd ed.) New York: McGraw-Hill Book Co., 1972.

26. Bearn, A. G., and H. G. Kunkel. Metabolic studies in Wilson's disease using Cu^{64}. J. Lab. Clin. Med. 45:623-631, 1955.

27. Beck, A. B., and H. W. Bennetts. Copper poisoning in sheep in Western Australia. J. Roy. Soc. West. Austral. 46:5-10, 1963.

28. Beeson, K. C. The Mineral Composition of Crops with Particular Reference to the Soils in Which They Were Grown. A Review and Compilation. U.S. Department of Agriculture Miscellaneous Publication 369. Washington, D.C.: U.S. Government Printing Office, 1941. 164 pp.

29. Benko, L., I. Beseda, and J. Galad. Cases of chronic copper poisoning in calves. Veterinarstvi 22:495-496, 1972. (in Russian)

30. Bennetts, H. W., A. B. Beck, and R. Harley. The pathogenesis of "falling disease." Austral. Vet. J. 24:237-244, 1948.

31. Bennetts, H. W., and H. T. B. Hall. "Falling disease" of cattle in the south-west of Western Australia. Austral. Vet. J. 15:152-159, 1939.

32. Bennetts, H. W., R. Harley, and S. T. Evans. Studies on copper deficiency of cattle: The fatal termination ("falling disease"). Austral. Vet. J. 18:50-63, 1942.

33. Berger, K. C. Micronutrient deficiencies in the United States. J. Agric. Food Chem. 10:178-181, 1962.

34. Berman, E. Biochemical applications of flame emission and atomic absorption spectroscopy. Appl. Spectrosc. 29:1-9, 1975.

35. Bickel, H., F. C. Neale, and G. Hall. A clinical and biochemical study of hepatolenticular degeneration (Wilson's disease). Q. J. Med. 26:527–558, 1957.
36. Bishop, N. I. Photosynthesis: The electron transport system of green plants. Ann. Rev. Biochem. 40:197–226, 1971.
37. Blaschko, H. Amine oxidase, pp. 337–351. In P. D. Boyer, H. Lardy, and K. Myrbäck, Eds. The Enzymes. Vol. 8. Oxidation and Reduction (Part B), Metal-Porphyrin Enzymes, Other Oxidases, Oxygenation, Topical Subject Index: Volumes 1–8. (2nd ed.) New York: Academic Press, 1963.
38. Blomfield, J. Copper contamination in exchange transfusions. Lancet 1:731–732, 1969.
39. Blomfield, J., S. R. Dixon, and D. A. McCredie. Potential hepatotoxicity of copper in recurrent hemodialysis. Arch. Intern. Med. 128:555–560, 1971.
40. Blomfield, J., J. McPherson, and C. R. P. George. Active uptake of copper and zinc during haemodialysis. Brit. Med. J. 2:141–145, 1969.
41. Bloomer, L. C., and T. L. Sourkes. The effect of copper loading on the distribution of copper in rat liver cytosol. Biochem. Med. 8:78–91, 1973.
42. Bohré, G. F., J. Huisman, and H. F. L. Lifferink. Acute copper poisoning aboard a ship. Ned. Tijdschr. Geneeskd. 109:978–979, 1965. (in Dutch)
43. Bortels, H. Über die Bedeutung von Eisen, Zink und Kupfer für Mikroorganismen. Biochem. Z. 182:301–358, 1927.
44. Bouchilloux, S., P. McMahill, and H. S. Mason. The multiple forms of mushroom tyrosinase. Purification and molecular properties of the enzymes. J. Biol. Chem. 238:1699–1707, 1963.
45. Boughton, I. B., and W. T. Hardy. Chronic Copper Poisoning in Sheep. Texas Agricultural Experiment Station Bulletin No. 499. College Station, Tex.: Agricultural and Mechanical College of Texas, 1934. 32 pp.
46. Boulard, M., K.-G. Blume, and E. Beutler. The effect of copper on red cell enzyme activities. J. Clin. Invest. 51:459–461, 1972.
47. Bowen, T. E., and T. W. Sullivan. Influence of dietary cupric sulfate on the response of young turkeys to penicillin-streptomycin (1:3). Poult. Sci. 50:273–278, 1971.
48. Bowler, K., and C. J. Duncan. The effect of copper on membrane enzymes. Biochim. Biophys. Acta 196:116–119, 1970.
49. Bowness, J. M., and R. A. Morton. Distribution of copper and zinc in the eyes of freshwater fishes and frogs. Occurrence of metals in melanin fractions from eye tissues. Biochem. J. 51:530–535, 1952.
50. Bowness, J. M., R. A. Morton, M. H. Shakir, and A. L. Stubbs. Distribution of copper and zinc in mammalian eyes. Occurrence of metals in melanin fractions from eye tissues. Biochem. J. 51:521–530, 1952.
51. Boyden, R., V. R. Potter, and C. A. Elvehjem. Effect of feeding high levels of copper to albino rats. J. Nutr. 15:397–402, 1938.
52. Bracewell, C. D. A note on jaundice in housed sheep. Vet. Rec. 70:342–343, 1958.
53. Brady, F. O., M. E. Monaco, H. J. Forman, G. Schutz, and P. Feigelson. On the role of copper in activation of and catalysis by tryptophan-2,3-dioxygenase. J. Biol. Chem. 247:7915–7922, 1972.
54. Braude, R., K. G. Mitchell, and R. J. Pittman. A note on cuprous chloride as a feed additive for growing pigs. Anim. Prod. 17:321–323, 1973.
55. Bray, R. C. Xanthine oxidase, pp. 533–556. In P. D. Boyer, H. Lardy and K. Myrbäck, Eds. The Enzymes. Vol. 7. Oxidation and Reduction (Part A),

Nicotinamide Nucleotide-Linked Enzymes, Flavin Nucleotide-Linked Enzymes. (2nd ed.) New York: Academic Press, 1963.

56. Bremner, I., and R. B. Marshall. Hepatic copper- and zinc-binding proteins in ruminants. Brit. J. Nutr. 32:293-330, 1974.

57. Breslow, E. Comparison of cupric ion-binding sites in myoglobin derivatives and serum albumin. J. Biol. Chem. 239:3252-3259, 1964.

58. Britton, J. W., and H. Goss. Chronic molybdenum poisoning in cattle. J. Amer. Vet. Med. Assoc. 108:176-178, 1946.

59. Broadbent, F. E., and T. Nakashima. The effect of added salts on nitrogen mineralization in three California soils. Soil Sci. Soc. Amer. Proc. 35:457-460, 1971.

60. Brodsky, J. Der Einfluss der kohlensauren Kupferbeizen für Getreide auf den Tierorganismus. Arch. Gewerbepath. Gewerbehyg. 5:91-107, 1933.

61. Broman, L. Chromatographic and magnetic studies on human ceruloplasmin. Acta Soc. Med. Upsal. 69(Suppl. 7):1-85, 1964.

62. Broman, L., B. G. Malmström, R. Aasa, and T. Vänngård. Quantitative electron spin resonance studies on native and denatured ceruloplasmin and laccase. J. Mol. Biol. 5:301-310, 1962.

63. Broman, L., B. G. Malmström, R. Aasa, and T. Vänngård. The role of copper in the catalytic action of laccase and ceruloplasmin. Biochim. Biophys. Acta 75:365-376, 1963.

64. Brooks, D. W., and C. R. Dawson. Aspects of tyrosinase chemistry, pp. 343-357. In J. Peisach, P. Aisen, and W. E. Blumberg, Eds. The Biochemistry of Copper. Proceedings of the Symposium on Copper in Biological Systems held at Arden House, Harriman, New York, September 8-10, 1965. New York: Academic Press, 1966.

65. Brown, F. C., and D. N. Ward. Studies on mammalian tyrosinase. II. Chemical and physical properties of fractions purified by chromatography. Proc. Soc. Exp. Biol. Med. 100:701-704, 1959.

66. Brown, J. C. Iron chlorosis in plants. Adv. Agron. 13:329-369, 1961.

67. Brown, V. M., and R. A. Dalton. The acute lethal toxicity to rainbow trout of mixtures of copper, phenol, zinc, and nickel. J. Fish. Biol. 2:211-216, 1970.

68. Browning, E. Toxicity of Industrial Metals. London: Butterworths, 1961. 325 pp.

69. Buchauer, M. J. Contamination of soil and vegetation near a zinc smelter by zinc, cadmium, copper and lead. Environ. Sci. Technol. 7:131-135, 1973.

70. Buck, W. B. Diagnosis of feed-related toxicoses. J. Amer. Vet. Med. Assoc. 156:1434-1443, 1970.

71. Buck, W. B., G. D. Osweiler, and G. A. Van Gelder. Clinical and Diagnostic Veterinary Toxicology. Dubuque, Iowa: Kendall/Hunt Publishing Company, 1973. 287 pp.

72. Buck, W. B., and R. M. Sharma. Copper toxicity in sheep. Iowa State Univ. Vet. 31:4-8, 1969.

73. Bull, L. B., H. E. Albiston, G. Edgar, and A. T. Dick. Toxaemic jaundice of sheep: Phytogenous chronic copper poisoning, heliotrope poisoning, and hepatogenous chronic copper poisoning. Final report of the investigation committee. Austral. Vet. J. 32:220-236, 1956.

74. Bunch, R. J., J. T. McCall, V. C. Speer, and V. W. Hays. Copper supplementation for weanling pigs. J. Anim. Sci. 24:995-1000, 1965.

75. Bunch, R. J., V. C. Speer, V. W. Hays, and J. T. McCall. Effects of high levels of

copper and chlortetracycline on performance of pigs. J. Anim. Sci. 22:56–60, 1963.

76. Butcher, L. L., and S. S. Fox. Motor effects of copper in the caudate nucleus: Reversible lesions with ion-exchange resin beads. Science 160:1237–1239, 1968.

77. Butler, E. J., and G. E. Newman. The urinary excretion of copper and its concentration in the blood of normal human adults. J. Clin. Path. 9:157–161, 1956.

78. Butterworth, C. E., Jr., C. J. Gubler, G. E. Cartwright, and M. M. Wintrobe. Studies on copper metabolism. XXVI. Plasma copper in patients with tropical sprue. Proc. Soc. Exp. Biol. Med. 98:594–597, 1958.

79. Cairns, J. E., H. P. Williams, and J. M. Walshe. "Sunflower cataract" in Wilson's disease. Brit. Med. J. 3:95–96, 1969.

80. Cannon, H. L., and B. M. Anderson. The geochemist's involvement with pollution problems, pp. 155–177. In H. L. Cannon and H. C. Hopps, Eds. Environmental Geochemistry in Health and Disease. American Association for the Advancement of Science Symposium, Dallas, Texas, December 1968. Memoir 123. Boulder, Colorado: Geological Society of America, 1971.

81. Carrico, R. J., and H. F. Deutsch. Isolation of human hepatocuprein and cerebrocuprein. Their identity with erythrocuprein. J. Biol. Chem. 244:6087–6093, 1969.

82. Carrico, R. J., and H. F. Deutsch. The presence of zinc in human cytocuprein and some properties of the apoprotein. J. Biol. Chem. 245:723–727, 1970.

83. Carrico, R. J., H. F. Deutsch, H. Beinert, and W. H. Orme-Johnson. Some properties of an apoceruloplasmin-like protein in human serum. J. Biol. Chem. 244:4141–4146, 1969.

84. Carruthers, M. E., C. B. Hobbs, and R. L. Warren. Raised serum copper and caeruloplasmin levels in subjects taking oral contraceptives. J. Clin. Path. 19:498–503, 1966.

85. Cartwright, G. E., C. J. Gubler, J. A. Bush, and M. M. Wintrobe. Studies on copper metabolism. XVII. Further observations on the anemia of copper deficiency in swine. Blood 11:143–153, 1956.

86. Cartwright, G. E., R. E. Hodges, C. J. Gubler, J. P. Mahoney, K. Daum, M. M. Wintrobe, and W. B. Bean. Studies on copper metabolism. XIII. Hepatolenticular degeneration. J. Clin. Invest. 33:1487–1501, 1954.

87. Cartwright, G. E., and M. M. Wintrobe. Copper metabolism in normal subjects. Amer. J. Clin. Nutr. 14:224–232, 1964.

88. Cartwright, G. E., and M. M. Wintrobe. The question of copper deficiency in man. Amer. J. Clin. Nutr. 15:94–110, 1964.

89. Chang, H. T. Further Studies on Apoascorbate Oxidase. Ph.D. Thesis. New York: Columbia University, 1970. 85 pp.

90. Chapman, H. L., Jr., S. L. Nelson, R. W. Kidder, W. L. Sippel, and C. W. Kidder. Toxicity of cupric sulfate for beef cattle. J. Anim. Sci. 21:960–962, 1962.

91. Chase, M. S., C. J. Gubler, G. E. Cartwright, and M. M. Wintrobe. Studies on copper metabolism. IV. The influence of copper on the absorption of iron. J. Biol. Chem. 199:757–763, 1952.

92. Chase, M. S., C. J. Gubler, G. E. Cartwright, and M. M. Wintrobe. Studies on copper metabolism. V. Storage of iron in liver of copper-deficient rats. Proc. Soc. Exp. Biol. Med. 80:749–751, 1952.

93. Chemodanova, E. I. The effect of molybdenum on the vitamin C content and ascorbic acid oxidase activity of leguminous plants, pp. 287–290. In I. I. Matusis,

Ed. Problems of Vitaminology. Barnaul, R.S.F.S.R.: Altaiskii Gosudarstuennyi Meditsinskii Institut, 1959. (in Russian)

94. Chiarandini, D. J., E. Stefani, and H. M. Gerschenfeld. Inhibition of membrane permeability to chloride by copper in molluscan neurones. Nature 213:97–99, 1967.

95. Chuttani, H. K., P. S. Gupta, S. Gulati, and D. N. Gupta. Acute copper sulfate poisoning. Amer. J. Med. 39:849–854, 1965.

96. Clawson, W. J., A. L. Lesperance, V. R. Bohman, and D. C. Layhee. Interrelationship of dietary molybdenum and copper on growth and tissue composition of cattle. J. Anim. Sci. 34:516–520, 1972.

97. Cohen, S. R. A review of the health hazards of copper exposure. J. Occup. Med. 16:621–624, 1974.

98. Compère, R., A. Burny, A. Riga, E. Francois, and S. Vanuytrecht. Copper in the treatment of molybdenosis in the rat: Determination of the toxicity of the antidote. J. Nutr. 87:412–418, 1965.

99. Copper. World Metal Statist. 27(6):33–66, 1974.

100. Copper shows more arthritis benefits. Chem. Eng. News 53(16):36–37, 1975. (abstract)

101. Cordano, A. Copper requirements and actual recommendations per 100 kilocalories of infant formula. Pediatrics 54:524, 1974. (letter)

102. Cordano, A., and G. G. Graham. Copper deficiency complicating severe chronic intestinal malabsorption. Pediatrics 38:596–604, 1966.

103. Cordano, A., R. P. Placko, and G. G. Graham. Hypocupremia and neutropenia in copper deficiency. Blood 28:280–283, 1966.

104. Cordy, D. R. Enzootic ataxia in California lambs. J. Amer. Vet. Med. Assoc. 158:1940–1942, 1971.

105. Cromwell, G. L. Copper, molybdenum, sulfate and sulfide interrelationships in swine. Anim. Nutr. Health 26(12):5–7, 1971.

106. Cuadros, A., and J. G. Hirsch. Copper on intrauterine devices stimulates leukocyte exudation. Science 175:175–176, 1972.

107. Cuffe, S. T., and R. W. Gerstle. Emissions from Coal-Fired Power Plants. A Comprehensive Summary. Public Health Publ. 999-AP-35. Washington, D.C.: U.S. Government Printing Office, 1967. 26 pp.

108. Cumings, J. N. Copper. Hepatolenticular degeneration, pp. 3–71. In Heavy Metals and the Brain. Oxford: Blackwell Scientific Publications Ltd., 1959.

109. Cumings, J. N. The effects of B.A.L. in hepatolenticular degeneration. Brain 74:10–22, 1951.

110. Cunha, T. J. Effect of antibiotic feeding, p. 159. In Swine Feeding and Nutrition. New York: Interscience Publishers, Inc., 1957.

111. Cunningham, I. J. Some biochemical and physiological aspects of copper in animal nutrition. Biochem. J. 25:1267–1294, 1931.

112. Cunningham, I. J., K. G. Hogan, and B. M. Lawson. The effect of sulfate and molybdenum on copper metabolism in cattle. N. Z. J. Agric. Res. 2:145–152, 1959.

113. Curzon, G. Some properties of coupled iron-caeruloplasmin oxidation systems. Biochem. J. 79:656–663, 1961.

114. Dale, S. E. Effect of Molybdenum and Sulfate on Copper Metabolism in Young Growing Pigs. M.S. Thesis. Ames: Iowa State University, 1971. 69 pp.

115. Dallman, P. R. Cytochrome oxidase repair during treatment of copper deficiency: Relation to mitochondrial turnover. J. Clin. Invest. 46:1819–1827, 1967.

116. D'Amico, K. J. Mineral production in the United States, pp. 4-5. In Minerals Yearbook. 1958. III. Area Reports. Washington, D.C.: U.S. Department of the Interior, Bureau of Mines, 1959.

117. Danks, D. M., E. Cartwright, B. J. Stevens, and R. R. W. Townley. Menkes' kinky hair disease: Further definition of the defect in copper transport. Science 179: 1140-1142, 1973.

118. Danks, D. M., B. J. Stevens, P. E. Campbell, J. M. Gillespie, J. Walker-Smith, J. Blomfield, and B. Turner. Menkes' kinky-hair syndrome. Lancet 1:1100-1102, 1972.

119. Davenport, S. J. Review of Literature on Health Hazards of Metals. I. Copper. U.S. Bureau of Mines Information Circular No. 7666. Washington, D.C.: U.S. Government Printing Office, 1953. 114 pp.

120. Davis, G. K. High-level copper feeding of swine and poultry and the ecology. Fed. Proc. 33:1194-1196, 1974.

121. Davis, G. K. The influence of copper on the metabolism of phosphorus and molybdenum, pp. 216-229. In W. D. McElroy and B. Glass, Eds. Copper Metabolism. A Symposium on Animal, Plant and Soil Relationships. Baltimore: The Johns Hopkins Press, 1950.

122. Davis, W. E., & Associates. National Inventory of Sources and Emissions. Barium, Boron, Copper, Selenium, and Zinc. 1969. Copper. Section III. Leawood, Kansas: W. E. Davis & Associates, 1972. 68 pp.

123. Dawson, C. R. Ascorbate oxidase, a review, pp. 305-337. In J. Peisach, P. Aisen, and W. E. Blumberg, Eds. The Biochemistry of Copper. Proceedings of the Symposium on Copper in Biological Systems held at Arden House, Harriman, New York, September 8-10, 1965.

124. Dawson, C. R. The copper protein, ascorbic acid oxidase, pp. 18-47. In W. D. McElroy and B. Glass, Eds. Copper Metabolism. A Symposium on Animal, Plant and Soil Relationships. Baltimore: Johns-Hopkins Press, 1950.

125. Dawson, C. R. The copper protein, ascorbic acid oxidase. Part II. The biological significance of chelation. Ann. N.Y. Acad. Sci. 88:353-360, 1960.

126. Dawson, C. R., and M. F. Mallette. Ascorbic acid oxidase. Adv. Protein Chem. 2:224-229, 1945.

127. Dawson, C. R., and W. B. Tarpley. Ascorbic acid oxidase, pp. 491-498. In J. B. Sumner and K. Myrbäck, Eds. The Enzymes. Vol. 2, Part 1. New York: Academic Press, Inc., 1951.

128. Deane, R. S., E. L. Mills, and A. J. Hamel. Antibacterial action of copper in respiratory therapy apparatus. Chest 58:373-377, 1970.

129. DeGoey, L. W., R. C. Wahlstrom, and R. J. Emerick. Studies of high level copper supplementation to rations for growing swine. J. Anim. Sci. 33:52057, 1971.

130. Deiss, A., G. R. Lee, and G. E. Cartwright. Hemolytic anemia in Wilson's disease. Ann. Intern. Med. 73:413-418, 1970.

131. Deiss, A., R. E. Lynch, G. R. Lee, and G. E. Cartwright. Long-term therapy of Wilson's disease. Ann. Intern. Med. 75:57-65, 1971.

132. de Jorge, F. B., H. M. Canelas, J. C. Dias, and L. Cury. Studies on copper metabolism. III. Copper contents of saliva of normal subjects and of salivary glands and pancreas of autopsy material. Clin. Chim. Acta 9:148-150, 1964.

133. Delas, J. La toxicité du cuivre accumulé dans les sols. Agrochimica 7:258-288, 1963.

134. Denny-Brown, D. Hepatolenticular degeneration (Wilson's disease). Two different components. New Engl. J. Med. 270:1149-1156, 1964.

135. Denny-Brown, D., and H. Porter. The effect of BAL (2,3-dimercaptopropanol) on hepatolenticular degeneration (Wilson's disease). New Engl. J. Med. 245:917–925, 1951.

136. Deschiens, R. Le contrôl de l'action des molluscicides chimiques sur les associations zoophytiques des eaux douces. C. R. Acad. Sci. (Paris) D 266:1860–1861, 1968.

137. Dick, A. T. Preliminary observations on the effect of high intakes of molybdenum and of inorganic sulfate on blood copper and on fleece character in crossbred sheep. Austral. Vet. J. 30:196–202, 1954.

138. Dick, A. T. The control of copper storage in the liver of sheep by inorganic sulfate and molybdenum. Austral. Vet. J. 29:233–239, 1953.

139. Dick, A. T. The effect of inorganic sulfate on the excretion of molybdenum in sheep. Austral. Vet. J. 29:18–26, 1953.

140. Dick, A. T., and L. B. Bull. Some preliminary observations on the effect of molybdenum on copper metabolism in herbivorous animals. Austral. Vet. J. 21:70–72, 1945.

141. Diem, K., and C. Lentner, Eds. Composition of foods, pp. 499–515. In Documenta Geigy. Scientific Tables. (7th ed.) Ardsley, N.Y.: Geigy Pharmaceuticals, Division of Ciba-Geigy Corporation, 1970.

142. Doherty, P. C., R. M. Barlow, and K. W. Angus. Spongy changes in the brains of sheep poisoned by excess dietary copper. Res. Vet. Sci. 10:303–304, 1969.

143. Doner, H. E., and M. M. Mortland. Benzene complexes with copper(II)-montmorillonite. Science 166:1406–1408, 1969.

144. Dowdy, R. P., G. A. Kunz, and H. E. Sauberlich. Effect of a copper-molybdenum compound upon copper metabolism in the rat. J. Nutr. 99:491–496, 1969.

145. Dowdy, R. P., and G. Matrone. A copper-molybdenum complex: Its effects and movement in the piglet and sheep. J. Nutr. 95:197–201, 1968.

146. Dowdy, R. P., and G. Matrone. Copper-molybdenum interaction in sheep and chicks. J. Nutr. 95:191–196, 1968.

147. Doudoroff, P., and M. Katz. Critical review of the literature on the toxicity of industrial wastes and their components to fish. II. The metals, as salts. Sew. Ind. Wastes 25:802–839, 1953.

148. Eden, A., and H. H. Green. Micro-determination of copper in biological material. Biochem. J. 34:1202–1208, 1940.

149. Ehrenberg, A., B. G. Malmström, L. Broman, and R. Mosbach. A magnetic susceptibility of copper valence in ceruloplasmin and laccase. J. Mol. Biol. 5:450–452, 1962.

150. Elliot, J. I., and J. P. Bowland. Effects of dietary copper sulfate and protein on the fatty acid composition of porcine fat. J. Anim. Sci. 30:923–930, 1970.

151. Engel, R. W., N. O. Price, and R. F. Miller. Copper, manganese, cobalt and molybdenum balance in pre-adolescent girls. J. Nutr. 92:197–204, 1967.

152. Epstein, E. Mineral Nutrition of Plants: Principles and Perspectives. New York: John Wiley & Sons, 1972. 412 pp.

153. Epstein, P. S., and H. McIlwain. Actions of cupric salts on isolated cerebral tissues. Proc. Roy. Soc. London B 166:295–302, 1966.

154. Erecińska, M., and D. F. Wilson. Kinetic studies on cytochrome b-c_1 interaction in the isolated succinate-cytochrome c reductase. FEBS Lett. 24:269–272, 1972.

155. Erickson, S. J., N. Lackie, and T. E. Maloney. A screening technique for estimating copper toxicity to estuarine phytoplankton. J. Water Pollut. Control Fed. 42(Res. Suppl.):R270–R278, 1970.

156. Evans, G. W. Copper homeostasis in the mammalian system. Physiol. Rev. 53:535–570, 1973.
157. Evans, G. W., R. S. Dubois, and K. M. Hambidge. Wilson's disease: Identification of an abnormal copper-binding protein. Science 181:1175–1176, 1973.
158. Evans, G. W., C. Hahn, and H. H. Sandstead. Copper- and zinc-binding components in rat intestine. Clin. Res. 21:867, 1973. (abstract)
159. Fairbanks, V. F. Copper sulfate-induced hemolytic anemia. Arch. Intern. Med. 120:428–432, 1967.
160. Fare, G. The protective effects of beef and yeast extracts and copper acetate in the diet against rat liver carcinogenesis by 4-dimethyl-aminoazobenzene. Brit. J. Cancer 18:782–791, 1964.
161. Federal Water Pollution Control Administration. Water Quality Criteria. Report of the National Technical Advisory Committee to the Secretary of the Interior April 1, 1968. Washington, D.C.: U.S. Government Printing Office, 1968. 234 pp.
162. Fee, J. A., R. Malkin, B. G. Malmström, and T. Vänngård. Anaerobic oxidation-reduction titrations of fungal laccase. Evidence for several high potential electron-accepting sites. J. Biol. Chem. 244:4200–4207, 1969.
163. Feenstra, P., and F. W. van Ulsen. Hay as a cause of copper poisoning in sheep. Tijdschr. Diergeneeskd. 98:632–633, 1973. (in Dutch)
164. Feig, S. A., G. B. Segel, S. B. Shohet, and D. G. Nathan. Energy metabolism in human erythrocytes. II. Effects of glucose depletion. J. Clin. Invest. 51:1547–1554, 1972.
165. Feldmann, G., C. Abramowitz, H. Sarmini, and F. Rousselet. Répartition du cuivre dans les fractions subcellulaires hépatiques après intoxication subaiguë par le cuivre chez le rat. Biol. Gastroenterol. 5:35–46, 1972.
166. Feldmann, G., O. Groussard, and R. Fauvert. L'ultrastructure hépatique au cours de la maladie de Wilson. Biol. Gastroenterol. 2:137–160, 1969.
167. Fell, G. S., H. Smith, and R. A. Howie. Neutron activation analysis for copper in biological material applied to Wilson's disease. J. Clin. Path. 21:8–11, 1968.
168. Ferguson, W. S., A. H. Lewis, and S. J. Watson. Action of molybdenum in nutrition of milking cattle. Nature 141:553, 1938.
169. Fiscina, B., G. K. Oster, G. Oster, and J. Swanson. Gonococcicidal action of copper in vitro. Amer. J. Obstet. Gynecol. 116:86–90, 1973.
170. Fishburn, C. W., and C. Zenz. Metal fume fever. A report of a case. J. Occup. Med. 11:142–144, 1969.
171. Fitzpatrick, T. B., and W. C. Quevedo, Jr. Albinism, pp. 326–337. In J. B. Stanbury, J. B. Wyngaarden and D. S. Fredrickson, Eds. The Metabolic Basis of Inherited Disease. (3rd ed.) New York: McGraw Hill, 1972.
172. Fitzpatrick, T. B., M. Seiji, and A. D. McGugan. Melanin pigmentation. New Engl. J. Med. 265:328–332, 1961.
173. Fitzpatrick, T. B., M. Seiji, and A. D. McGugan. Melanin pigmentation (continued). New Engl. J. Med. 265:374–378, 1961.
174. Fitzpatrick, T. B., M. Seiji, and A. D. McGugan. Melanin Pigmentation (concluded). New Engl. J. Med. 265:430–434, 1961.
175. Fleming, C. E., J. A. McCormick, and W. B. Dye. The Effects of Molybdenosis on a Growth and Breeding Experiment. Nevada Agricultural Experiment Station Bulletin No. 220. Reno: University of Nevada, 1961. 15 pp.
176. Franke, W. Phenoloxydasen und Ascorbinsäureoxydase, pp. 401–455. In W. Ruhland, Ed. Handbuch der Pflanzenphysiologie. Vol. 12/Part 1. Pflanzenatmung Einschliesslich Gärungen und Säurestoffwechsel. Berlin: Springer-Verlag, 1960.

177. French, J. H., C. L. Moore, N. R. Ghatak, I. Sternlieb, S. Goldfischer, and
      A. Hirano. Trichopoliodystrophy (Menkes' kinky hair syndrome): A copper
      dependent deficiency of mitochondrial energetics. Pediatr. Res. 7:386–387,
      1973. (abstract)

178. French, J. H., E. S. Sherard, H. Lubell, M. Brotz, and C. L. Moore. Trichopolio-
      dystrophy. I. Report of a case and biochemical studies. Arch. Neurol. 26:229–
      244, 1972.

179. Frenyo, V., and Thai Duy Ninh. Effect of copper ions on gas evolution during
      photosynthesis. Bot. Közlem. 57:107–112, 1970. (in Hungarian, summary in
      German)

180. Freudenberg, K. Biosynthesis and constitution of lignin. Nature 183:1152–1155,
      1959.

181. Fridovich, I. Superoxide dismutases. Adv. Enzymol. 41:35–97, 1974.

182. Friedman, S., and S. Kaufman. 3,4-dihydroxyphenylethylamine β-hydroxylase:
      A copper protein. J. Biol. Chem. 240:552–554, 1965.

183. Frommer, D. J. The measurement of biliary copper secretion in humans. Clin.
      Sci. 42:26P, 1972. (abstract)

184. Frykholm, K. O., L. Frithiof, A. I. B. Fernström, G. Moberger, S. G. Blohm, and
      E. Björn. Allergy to copper derived from dental alloys as a possible cause of oral
      lesions of lichen planus. Acta Derm. Venereol. 49:268–281, 1969.

185. Funderburk, H. H., Jr., and D. E. Davis. Metabolism of C^{14} chain- and ring-
      labeled simazine by corn and the effect of atrazine on plant respiratory systems.
      Weeds 11:101–104, 1963.

186. Fushimi, H., C. R. Hamison, and H. A. Ravin. Two new copper proteins from
      human brain. Isolation and properties. J. Biochem. (Tokyo) 69:1041–1054,
      1971.

187. Gallagher, C. H., and V. E. Reeve. Copper deficiency in the rat. Effect on syn-
      thesis of phospholipids. Austral. J. Exp. Biol. Med. Sci. 49:21–31, 1971.

188. Gardiner, M. R. Mineral metabolism in sheep lupinosis. II. Copper. J. Comp.
      Path. 76:107–120, 1966.

189. Gardiner, M. R. The role of copper in the pathogenesis of subacute and chronic
      lupinosis of sheep. Austral. Vet. J. 43:243–248, 1967.

190. Gerhard, J.-P. Étude du cuivre de l'humeur aqueuse. Doc. Ophthalmol. 20:104–
      110, 1966.

191. Gerlach, W. Über den Kupfergehalt menschlicher Tumoren in Beziehung zum
      Kupfergehalt der Leber. Z. Krebsforsch. 42:290–294, 1935.

192. Gilsanz, V., A. Barrera, and A. Anaya. The renal biopsy in Wilson's disease.
      A.M.A. Arch. Intern. Med. 105:758–761, 1960.

193. Gipp, W. F., W. G. Pond, F. A. Kallfelz, J. B. Tasker, D. R. Van Campen,
      L. Krook, and W. J. Visek. Effect of dietary copper, iron, and ascorbic acid
      levels on hematology, blood and tissue copper, iron, and zinc concentrations
      and ^{64}Cu and ^{59}Fe metabolism in young pigs. J. Nutr. 104:532–541, 1974.

194. Gipp, W. F., W. G. Pond, and S. E. Smith. Effects of level of dietary copper,
      molybdenum, sulfate and zinc on bodyweight gain, hemoglobin, and liver
      copper storage of growing pigs. J. Anim. Sci. 26:727–730, 1967.

195. Gleason, R. P. Exposure to copper dust. Amer. Ind. Hyg. Assoc. J. 29:461–462,
      1968.

196. Goldberg, A., C. B. Williams, R. S. Jones, M. Yamagita, G. E. Cartwright, and
      M. M. Wintrobe. Studies on copper metabolism. XXII. Hemolytic anemia in

chickens induced by the administration of copper. J. Lab. Clin. Med. 48:442–453, 1956.

197. Goldfischer, S. Demonstration of copper and acid phosphatase activity in hepatocyte lysosomes in experimental copper toxicity. Nature 215:74–75, 1967.

198. Goldfischer, S., and J. Bernstein. Lipofuscin (aging) pigment granules of the newborn human liver. J. Cell Biol. 42:253–261, 1969.

199. Goldfischer, S., B. Schiller, and I. Sternlieb. Copper in hepatocyte lysosomes of the toad, *Bufo marinus L.* Nature 228:172–173, 1970.

200. Goldfischer, S., and I. Sternlieb. Changes in the distribution of hepatic copper in relation to the progression of Wilson's disease (hepatolenticular degeneration). Amer. J. Path. 53:883–901, 1968.

201. Goldstein, N. P., J. C. Ewert, R. V. Randall, and J. B. Gross. Psychiatric aspects of Wilson's disease (hepatolenticular degeneration): Results of psychometric tests during long-term therapy. Amer. J. Psychiatry 124:1555–1561, 1968.

202. Gollan, J. L., and D. J. Deller. Studies on the nature and excretion of biliary copper in man. Clin. Sci. 44:9–15, 1973.

203. Goodman, S. I., D. O. Rodgerson, and J. Kauffman. Hypercupremia in a patient with multiple myeloma. J. Lab. Clin. Med. 70:57–62, 1967.

204. Goodrich, R. D., and A. D. Tillman. Copper, sulfate, and molybdenum interrelationships in sheep. J. Nutr. 90:76–80, 1966.

205. Gopalachari, N. C. Changes in the activities of certain oxidizing enzymes during germination & seedling development of *Phaseolus mungo* & *Sorghum vulgare.* Indian J. Exp. Biol. 1:98–100, 1963.

206. Gorman, D. S., and R. P. Levine. Cytochrome *f* and plastocyanin: Their sequence in the photosynthetic electron transport chain of *Chlamydomonos reinhardi.* Proc. Nat. Acad. Sci. U.S.A. 54:1665–1669, 1965.

207. Gorman, D. S., and R. P. Levine. Photosynthetic electron transport chain of *Chlamydomonas reinhardi.* VI. Electron transport in mutant strains lacking either cytochrome 553 or plastocyanin. Plant Physiol. 41:1648–1656, 1966.

208. Grabske, R. J. The effect of Cu^{2+} on oxidative phosphorylation and the structural stability of isolated mitochondria. J. Cell Biol. 35:48A–49A, 1967. (abstract)

209. Graham, G. G., and A. Cordano. Copper depletion and deficiency in the malnourished infant. Johns Hopkins Med. J. 124:139–150, 1969.

210. Grant-Frost, D. R., and E. J. Underwood. Zinc toxicity in the rat and its interrelationship to copper. Austral. J. Exp. Biol. Med. Sci. 36:339–345, 1958.

211. Gray, I. F., and L. J. Daniel. Effect of the copper status of the rat on the copper-molybdenum-sulfate interaction. J. Nutr. 84:31–37, 1964.

212. Greenfield, S. S. Inhibitory effects of inorganic compounds on photosynthesis in *Chlorella.* Amer. J. Bot. 29:121–131, 1942.

213. Gregoriadis, G., A. G. Morell, I. Sternlieb, and I. H. Scheinberg. Catabolism of desialylated ceruloplasmin in the liver. J. Biol. Chem. 245:5833–5837, 1970.

214. Grover, W. D., and M. C. Scrutton. Copper infusion therapy in trichopoliodystrophy. J. Pediatr. 86:216–220, 1975.

215. Gubler, C. J., G. E. Cartwright, and M. M. Wintrobe. Studies on copper metabolism. XX. Enzyme activities and iron metabolism in copper and iron deficiencies. J. Biol. Chem. 224:533–546, 1957.

216. Gubler, C. J., M. E. Lahey, M. S. Chase, G. E. Cartwright, and M. M. Wintrobe. Studies on copper metabolism. III. The metabolism of iron in copper-deficient swine. Blood 7:1075–1092, 1952.

217. Gutmanis, K. Biosynthesis of riboflavine and enzymic processes in seeds. Latvijas PSR Zinatnu Akad. Vestis 1959(3):73-75. (in Latvian, summary in Russian)

218. Hadjimarkos, D. M. Effect of trace elements in drinking water on dental caries. J. Pediatr. 70:967-969, 1967.

219. Hagenfeldt, K. Intrauterine contraception with the copper-T device. 1. Effect on trace elements in the endometrium, cervical mucus and plasma. Contraception 6:37-54, 1972.

220. Hamilton, G. A., R. D. Libby, R. C. Hartzell. The valence of copper and the role of superoxide in the D-galactose oxidase catalyzed reaction. Biochem. Biophys. Res. Commun. 55:333-340, 1973.

221. Hampton, J. K., Jr., L. J. Rider, T. J. Goka, and J. P. Preslock. The histaminase activity of ceruloplasmin. Proc. Soc. Exp. Biol. Med. 141:974-977, 1972.

222. Hampton, R. E. Activity of some soluble oxidases in carrot slices infected with *Thielaviopsis basicola*. Phytopathology 53:497-499, 1963.

223. Hanna, C., and F. T. Fraunfelder. Lens capsule change after intraocular copper. Ann. Ophthalmol. 5:9-22, 1973.

224. Hanrahan, T. J., and J. F. O'Grady. Copper supplementation of pig diets. The effect of protein level and zinc supplementation on the response to added copper. Anim. Prod. 10:423-432, 1968.

225. Hardy, R. M., J. B. Stevens, and C. M. Stowe. Chronic progressive hepatitis in Bedlington terriers associated with elevated liver copper concentrations. Minnesota Vet. 15(2):13-24, 1975.

226. Harms, R. H., and D. P. Eberst. Influence of dietary cupric sulfate on the response of young turkeys to sodium sulfate. Poult. Sci. 53:1629-1631, 1974.

227. Hartz, J. W., S. Funakoshi, and H. F. Deutsch. The levels of superoxide dismutase and catalase in human tissues as determined immunochemically. Clin. Chim. Acta 46:125-132, 1973.

228. Hassall, K. A. Uptake of copper and its physiological effect on *Chlorella vulgaris*. Physiol. Plant. 16:323-332, 1963.

229. Hays, V. W., and R. D. Kline. Copper-molybdenum-sulfate interrelationships in growing pigs. Feedstuffs 41(44):18, 1969.

230. Hazel, C. R., and S. J. Meith. Bioassay of king salmon eggs and sac fry in copper solutions. Calif. Fish. Game 56:121-124, 1970.

231. Henkin, R. I., H. R. Keiser, I. A. Jaffe, I. Sternlieb, and I. H. Scheinberg. Decreased taste sensitivity after D-penicillamine, reversed by copper administration. Lancet 2:1268-1271, 1967.

232. Hernandez, O., L. M. Ballesteros, J. D. Mendez, and A. Rosado. Copper as a dissociating agent of liver and endometrial polysomes. Fertil. Steril. 25:108-112, 1974.

233. Herrmann, R., and W. Lang. Serum-Kupfer-Analysen mit Hilfe der Absorptions-Flammenphotometrie. Z. Klin. Chem. 1:182-186, 1963.

234. Hill, C. H., B. Starcher, and C. Kim. Role of copper in the formation of elastin. Fed. Proc. 26:129-133, 1967.

235. Hill, J. M. The changes with age in the distribution of copper and some copper-containing oxidases in red clover (*Trifolium pratense* L. cv. Dorset Marlgrass). Appendix—The determination of nanogram amounts of copper in plant materials using the copper-free apoenzyme of pea-seedling diamine oxidase. J. Exp. Bot. 24:525-536, 1973.

236. Hind, G. The site of action of plastocyanin in chloroplasts treated with detergent. Biochim. Biophys. Acta 153:235-240, 1968.

237. Hind, G., and J. M. Olson. Electron transport pathways in photosynthesis. Ann. Rev. Plant Physiol. 19:249–282, 1968.

238. Hoffman, G. L., and R. A. Duce. Copper contamination of atmospheric particulate samples collected with Gelman Hurricane samplers. Environ. Sci. Technol. 5:1134–1135, 1971.

239. Hogan, K. G., D. F. L. Money, and A. Blayney. The effect of a molybdate and sulphate supplement on the accumulation of copper in the livers of penned sheep. N. Z. J. Agric. Res. 11:435–444, 1968.

240. Hohnadel, D. C., F. W. Sunderman, Jr., M. W. Nechay, and M. D. McNeely. Atomic absorption spectrometry of nickel, copper, zinc, and lead in sweat collected from healthy subjects during sauna bathing. Clin. Chem. 19:1288–1292, 1973.

241. Holmberg, C. G., and C.-B. Laurell. Investigations in serum copper. III. Caeruloplasmin as an enzyme. Acta Chem. Scand. 5:476–480, 1951.

242. Holmberg, C. G., and C.-B. Laurell. Oxidase reactions in human plasma caused by caeruloplasmin. Scand. J. Clin. Lab. Invest. 3:103–107, 1951.

243. Holtzman, N. A., D. A. Elliott, and R. H. Heller. Copper intoxication. Report of a case with observations on ceruloplasmin. New Engl. J. Med. 275:347–352, 1966.

244. Holtzman, N. A., and B. M. Gaumnitz. Identification of an apoceruloplasmin-like substance in the plasma of copper-deficient rats. J. Biol. Chem. 245:2350–2353, 1970.

245. Holtzman, N. A., and R. H. A. Haslam. Elevation of serum copper following copper sulfate as an emetic. Pediatrics 42:189–193, 1968.

246. Hopper, S. H., and H. S. Adams. Copper poisoning from vending machines. Public Health Rep. 73:910–914, 1958.

247. Howell, J. S. The effect of copper acetate on p-dimethylamino-azobenzene carcinogenesis in the rat. Brit. J. Cancer 12:594–608, 1958.

248. Huber, J. T., N. O. Price, and R. W. Engel. Response of lactating dairy cows to high levels of dietary molybdenum. J. Anim. Sci. 32:364–367, 1971.

249. Huisingh, J., and G. Matrone. Copper-molybdenum interactions with the sulfate-reducing system in rumen microorganisms. Proc. Soc. Exp. Biol. Med. 139:518–521, 1972.

250. Hunt, A. H., R. M. Parr, D. M. Taylor, and N. G. Trott. Relation between cirrhosis and trace metal content of liver. With special reference to primary biliary cirrhosis and copper. Brit. Med. J. 2:1498–1501, 1963.

251. Hunt, C. E., J. Landesman, and P. M. Newberne. Copper deficiency in chicks: Effects of ascorbic acid on iron, copper, cytochrome oxidase activity, and aortic mucopolysaccharides. Brit. J. Nutr. 24:607–614, 1970.

252. Hunt, D. M. Primary defect in copper transport underlies mottled mutants in the mouse. Nature 249:852–854, 1974.

253. Iodice, A. A., D. A. Richer, and M. P. Schulman. Copper content of purified δ-amino-levulinic acid dehydrase. Fed. Proc. 17:248, 1958. (abstract)

254. Ishmael, J., and C. Gopinath. Blood copper and serum enzyme changes following copper calcium E.D.T.A. administration to hill sheep of low copper status. J. Comp. Path. 81:455–461, 1971.

255. Ishmael, J., and C. Gopinath. Effect of a single small dose of inorganic copper on the liver of sheep. J. Comp. Path. 82:47–57, 1972.

256. Ishmael, J., C. Gopinath, and J. M. Howell. Experimental chronic copper toxicity in sheep. Biochemical and haematological studies during the development of lesions in the liver. Res. Vet. Sci. 13:22–29, 1972.

257. Ishmael, J., C. Gopinath, and J. M. Howell. Experimental chronic copper toxicity

in sheep. Histological and histochemical changes during the development of lesions in the liver. Res. Vet. Sci. 12:358–366, 1971.

258. Ishmael, J., C. Gopinath, and J. M. Howell. Studies with copper calcium E.D.T.A. Acute toxicity in housed sheep. J. Comp. Path. 81:279–290, 1971.

259. Joester, K.-E., G. Jung, U. Weber, and U. Weser. Superoxide dismutase activity of Cu^{2+}-amino acid chelates. FEBS Lett. 25:25–28, 1972.

260. Jones, J. B., Jr. Plant tissue analysis for micronutrients, pp. 319–346. In J. J. Mortvedt, P. M. Giordano and W. L. Lindsay, Eds. Micronutrients in Agriculture. Proceedings of a Symposium, 1971. Madison, Wisc.: Soil Science Society of America, 1971.

261. Jones, J. R. E. The relation between electrolytic solution pressures of the metals and their toxicity to the stickleback (Gasterosteus aculeatus L.). J. Exp. Biol. 16:425–437, 1939.

262. Jones, J. R. E. The relative toxicity of salts of lead, zinc, and copper to the stickleback (Gasterosteus aculeatus L.) and the effect of calcium on the toxicity of lead and zinc salts. J. Exp. Biol. 15:394–407, 1938.

263. Josephs, H. W. Treatment of anaemia of infancy with iron and copper. Bull. Johns Hopkins Hosp. 49:246–258, 1931.

264. Jubb, K. V. F., and P. C. Kennedy. The haemopoietic system, pp. 297–406. In Pathology of Domestic Animals. Vol. 1. (2nd ed.) New York: Academic Press, 1970.

265. Kägi, J. H. R., S. R. Himmelhoch, P. D. Whanger, J. L. Bethune, and B. L. Vallee. Equine hepatic and renal metallothioneins. J. Biol. Chem. 249:3537–3542, 1974.

266. Kanabrocki, E. L., T. Fields, C. F. Decker, L. F. Case, E. B. Miller, E. Kaplan, and Y. T. Oester. Neutron activation studies of biological fluids: Manganese and copper. Int. J. Appl. Radiat. Isot. 15:175–190, 1964.

267. Karpel, J. T., and V. H. Peden. Copper deficiency in long-term parenteral nutrition. J. Pediatr. 80:32–36, 1972.

268. Kasper, C. B., and H. F. Deutsch. Immunochemical studies of crystalline human ceruloplasmin and derivatives. J. Biol. Chem. 238:2343–2350, 1963.

269. Kasper, C. B., and H. F. Deutsch. Physicochemical studies of human ceruloplasmin. J. Biol. Chem. 238:2325–2337, 1963.

270. Katoh, S., and A. San Pietro. Activities of chloroplast fragments. I. Hill reaction and ascorbate-indophenol photoreductions. J. Biol. Chem. 241:3575–3581, 1966.

271. Katoh, S., and A. San Pietro. The role of plastocyanin in NADP photoreduction by chloroplasts, pp. 407–422. In J. Peisach, P. Aisen, and W. E. Blumberg, Eds. The Biochemistry of Copper. Proceedings of the Symposium on Copper in Biological Systems held at Arden House, Harriman, New York, September 8–10, 1965. New York: Academic Press, 1966.

272. Katoh, S., I. Shiratori, and A. Takamiya. Purification and some properties of spinach plastocyanin. J. Biochem. (Tokyo) 51:32–40, 1962.

273. Katoh, S., I. Suga, I. Shiratori, and A. Takamiya. Distribution of plastocyanin in plants, with special reference to its localization in chloroplasts. Arch. Biochem. Biophys. 94:136–141, 1961.

274. Katoh, S., and A. Takamiya. Nature of copper-protein binding in spinach plastocyanin. J. Biochem. (Tokyo) 55:378–387, 1964.

275. Katoh, S., and A. Takamiya. Photochemical reactions of plastocyanin in chloroplasts, pp. 262–272. In National Research Council. Committee on Photobiology. Photosynthetic Mechanisms of Green Plants. NRC Publication 1145. Washington, D.C.: National Academy of Sciences, 1963.

276. Katoh, S., and A. Takamiya. The iron-protein binding in photosynthetic pyridine nucleotide reductase. Arch. Biochem. Biophys. 102:189–200, 1963.

277. Keilin, D., and T. Mann. Laccase, a blue copper-protein oxidase from the latex of *Rhus succedanea*. Nature 143:23–24, 1939.

278. Kellerman, K. F. The rational use of disinfectants and algicides in municipal water supplies, pp. 241–245. In Original Communications. Eighth International Congress of Applied Chemistry. Washington and New York, September 4 to 13, 1912. Vol. 26.

279. Kertész, D. State of copper in polyphenoloxidase (tyrosinase). Nature 180:506–507, 1957.

280. Kertész, D. The copper of polyphenoloxidase, pp. 359–369. In J. Peisach, P. Aisen, and W. E. Blumberg, Eds. The Biochemistry of Copper. Proceedings of the Symposium on Copper in Biological Systems held at Arden House, Harriman, New York September 8–10, 1965. New York: Academic Press, 1966.

281. Kertész, D., and R. Zito. Mushroom polyphenol oxidase. I. Purification and general properties. Biochim. Biophys. Acta 96:447–462, 1965.

282. Key, J. L. Changes in ascorbic acid metabolism associated with auxin-induced growth. Plant Physiol. 37:349–356, 1962.

283. Kick, H. Pflanzennährstoff, pp. 90–122. In H. Linser, Ed. Handbuch der Pflanzenernährung und Düngung. Vol. I/1. New York: Springer-Verlag, 1969.

284. Kimmel, J. R., H. Markowitz, and D. M. Brown. Some chemical and physical properties of erythrocuprein. J. Biol. Chem. 234:46–50, 1959.

285. Klein, L. A., M. Lang, N. Nash, and S. L. Kirschner. Sources of metals in New York City wastewater. J. Water Pollut. Control Fed. 46:2653–2662, 1974.

286. Klein, W. J., Jr., E. N. Metz, and A. R. Price. Acute copper intoxication. Arch. Intern. Med. 129:578–582, 1972.

287. Kline, R. K., V. W. Hays, and G. L. Cromwell. Effect of molybdenum and sulfate on copper status of pigs. J. Anim. Sci. 31:205, 1970. (abstract)

288. Kline, R. D., V. W. Hays, and G. L. Cromwell. Effects of copper, molybdenum and sulfate on performance, hematology and copper stores of pigs and lambs. J. Anim. Sci. 33:771–779, 1971.

289. Konovalov, N. V. Hepatocerebral Dystrophy. [Hepatolenticular Degeneration]. Moscow: Medgiz, 1960. 556 pp. (in Russian)

290. Kopp, J. F., and R. C. Kroner. Trace Metals in Waters of the United States. A Five Year Summary of Trace Metals in Rivers and Lakes of the United States (Oct. 1, 1962–Sept. 30, 1967) Cincinnati: U.S. Department of the Interior. Federal Water Pollution Control Administration, 1969. 212 pp.

291. Kovalenko, V. F. Effect of copper on the photosynthesis of apple leaves. Dokl. TSKhA 158:127–133, 1970. (in Russian)

292. Kowalczyk, T., A. L. Pope, K. C. Berger, and B. A. Muggenburg. Chronic copper toxicosis in sheep fed dry feed. J. Amer. Vet. Med. 145:352–357, 1964.

293. Kowalczyk, T., A. L. Pope, and D. K. Sorensen. Chronic copper poisoning in sheep resulting from free-choice, trace mineral-salt ingestion. J. Amer. Vet. Med. Assoc. 141:362–366, 1962.

294. Krul, K. G. Nature of Apo-ascorbate Oxidase, Its Reconstitution and Holoenzyme. Ph.D. Thesis. New York: Columbia University, 1973. 207 pp.

295. Kubota, J., and W. H. Allaway. Geographic distribution of trace element problems, pp. 525–553. In J. J. Mortvedt, P. M. Giordano, and W. L. Lindsay, Eds. Micronutrients in Agriculture. Proceedings of a Symposium held at Muscle Shoals, Alabama, April 20–22, 1971. Madison, Wis.: Soil Science Society of America, Inc., 1972.

296. Kubowitz, F. Spaltung und Resynthese der Polyphenoloxydase und des Hämocyanins. Biochem. Z. 299:32–57, 1938.

297. Kühn, H. Das Erkennen von Nährstoffmangelerscheinungen, pp. 992–1006. In

H. Linser, Ed. Handbuch der Pflanzenernährung und Düngung. Vol. I/2. New York: Springer-Verlag, 1972.

298. Kun, E., and D. S. Fanshier. Isolation and properties of a β-mercaptopyruvate-cleaving copper enzyme. Biochim. Biophys. Acta 32:338–348, 1959.

299. Kunath, B. Zur Psychopathologie der hepato-zerebralen Degeneration. Fortschr. Neurol. Psychiatr. 37:91–106, 1969.

300. Kuratsune, M., S. Tokudome, T. Shirakusa, M. Yoshida, Y. Tokumitsu, T. Hayano, and M. Seita. Occupational lung cancer among copper smelters. Int. J. Cancer 13:552–558, 1974.

301. Lahey, M. E., C. J. Gubler, M. S. Chase, G. E. Cartwright, and M. M. Wintrobe. Studies on copper metabolism. II. Hematologic manifestations of copper deficiency in swine. Blood 7:1053–1074, 1952.

302. Lailach, G. E., T. D. Thompson, and G. W. Brindley. Absorption of pyrimidines, purines and nucleosides by Co-, Ni-, Cu-, and Fe(III)-montmorillonite (clay-organic studies XIII). Clays Minerals 16:295–301, 1968.

303. Lal, S., and T. L. Sourkes. Deposition of copper in rat tissues—the effect of dose and duration of administration of copper sulfate. Toxicol. Appl. Pharmacol. 20:269–283, 1971.

304. Lal, S., and T. L. Sourkes. Intracellular distribution of copper in the liver during chronic administration of copper sulfate to the rat. Toxicol. Appl. Pharmacol. 18:562–572, 1971.

305. Lee, A. M., and J. F. Fraumeni, Jr. Arsenic and respiratory cancer in man: An occupational study. J. Nat. Cancer Inst. 42:1045–1052, 1969.

306. Lee, G. R., G. E. Cartwright, and M. M. Wintrobe. Heme biosynthesis in copper deficient swine. Proc. Soc. Exp. Biol. Med. 127:977–981, 1968.

307. Lee, G. R., S. Nacht, J. N. Lukens, and G. E. Cartwright. Iron metabolism in copper-deficient swine. J. Clin. Invest. 47:2058–2069, 1968.

308. Lee, M. H., and C. R. Dawson. Ascorbate oxidase. Further studies on the purification of the enzyme. J. Biol. Chem. 248:6596–6602, 1973.

309. Leeper, G. W. Reactions of Heavy Metals with Soils with Special Regard to Their Application in Sewage Wastes. Report under Contract DACW73-73-C-0026. Washington, D.C.: U.S. Department of the Army, Corps of Engineers, 1972. 70 pp.

310. Le Riche, H. H. Metal contamination of soil in the Woburn market-garden experiment resulting from the application of sewage sludge. J. Agric. Sci. (Cambridge) 71:205–208, 1968.

311. Leu, M. L., G. T. Strickland, W. M. Beckner, T. S. M. Chen, C. C. Wang, and S. J. Yeh. Muscle copper, zinc, and manganese levels in Wilson's disease: Studies with the use of neutron-activation analysis. J. Lab. Clin. Med. 76:432–438, 1971.

312. Likens, G. E., and F. H. Bormann. Acid rain: A serious environmental problem. Science 184:1176–1179, 1974.

313. Lindquist, R. R. Studies on the pathogenesis of hepatolenticular degeneration. I. Acid phosphatase activity in copper-loaded rat livers. Amer. J. Path. 51:471–481, 1967.

314. Lindquist, R. R. Studies on the pathogenesis of hepatolenticular degeneration. III. The effect of copper on rat liver lysosomes. Amer. J. Path. 53:903–927, 1968.

315. Lippes, J., M. Zielezny, and H. Sultz. The effect of copper on Loop A. J. Reprod. Med. 10:166–168, 1973.

316. Lorber, A. Nonceruloplasmin copper in rheumatoid arthritis. Arthr. Rheum. 12:459–460, 1969. (letter)

317. Lovett-Janison, P. L., and J. M. Nelson. Ascorbic acid oxidase from summer crook-neck squash (*C. pepo condensa*). J. Amer. Chem. Soc. 62:1409–1412, 1940.

318. Low-Copper Diet. West Point, Penn.: Merck Sharp & Dohme (not dated). 12 pp.

319. Lüke, F. Chronische Kupfervergiftungen und Spreicherung von Kupfer in den Lebern von Schafen verschiedener Rassen. Tierzüchter 23:283–284, 1971.

320. Lyle, W. H., J. E. Payton, and M. Hui. Haemodialysis and copper fever. Lancet 2:1324–1325, 1976.

321. Mache, R. Étude de la respiration et de l'activité de quelques oxydases de feuilles de Sarrasin (*Fagopyrum esculentum* M.) carencé ou non en bore. C. R. Acad. Sci. D (Paris) 256:1583–1585, 1963.

322. MacPherson, A., and R. G. Hemingway. The relative merit of various blood analysis and liver function tests in giving an early diagnosis of chronic copper poisoning in sheep. Brit. Vet. J. 125:213–221, 1969.

323. Magdoff-Fairchild, B., F. M. Lovell, and B. W. Low. An x-ray crystallographic study of ceruloplasmin. Determination of molecular weight. J. Biol. Chem. 244:3497–3499, 1969.

324. Mahler, H. R. Studies on the fatty acid oxidizing system of animal tissues. IV. The prosthetic group of butyryl coenzyme A dehydrogenase. J. Biol. Chem. 206:13–26, 1954.

325. Mäkisara, P., H.-M. Ruutsalo, M. Nissilä, A. Ruotsi, and G.-L. Mäkisara. Serum copper in rheumatoid arthritis and ankylosing spondylitis. Ann. Med. Exp. Biol. Fenn. 46:177–178, 1968.

326. Malek, E. A. A note on the use of copper compounds as molluscicides, pp. 171–175. In T. C. Cheng, Ed. Molluscicides in Schistosomiasis Control. New York: Academic Press, Inc., 1974.

327. Malkin, R., and B. G. Malmström. The state and function of copper in biological systems. Adv. Enzymol. 33:177–244, 1970.

328. Malkin, R., B. G. Malmström, and T. Vänngård. Spectroscopic differentiation of the electron-accepting sites in fungal laccase. Association of a near ultraviolet band with two electron-accepting unit. Eur. J. Biochem. 10:324–329, 1969.

329. Mallette, M. F., and C. R. Dawson. On the nature of highly purified mushroom tyrosinase preparations. Arch. Biochem. Biophys. 23:29–44, 1949.

330. Mallory, F. B. and F. Parker, Jr. Experimental copper poisoning. Amer. J. Path. 7:351–363, 1931.

331. Malmström, B. G., L. E. Andréasson, and B. Reinhammar. Copper-containing oxidases and superoxide dismutase, pp. 507–579. In P. D. Boyer, Ed. The Enzymes. Vol. 12. Oxidation-Reduction. Part B. Electron Transfer(II), Oxygenases, Oxidases(I). (3rd ed.) New York: Academic Press, 1975.

332. Malmström, B. G., A. Finazzi Agrò, and E. Antonini. The mechanism of laccase-catalyzed oxidations: Kinetic evidence for the involvement of several electron-accepting sites in the enzyme. Eur. J. Biochem. 9:383–291, 1969.

333. Malmström, B. G., B. Reinhammar, and T. Vänngård. Two forms of copper(II) in fungal laccase. Biochim. Biophys. Acta 156:67–76, 1968.

334. Mandelli, E. F. The inhibitory effects of copper on marine phytoplankton. Contrib. Mar. Sci. 14:47–57, 1969.

335. Mann, T., and D. Keilin. Haemocuprein and hepatocuprein, copper-protein compounds of blood and liver in mammals. Proc. Roy. Soc. London B 126:303–315, 1938.

336. Manzler, A. D., and A. W. Schreiner. Copper-induced acute hemolytic anemia. A new complication of hemodialysis. Ann. Intern. Med. 73:409–412, 1970.

337. Marcilese, N. A., C. B. Ammerman, R. M. Valsecchi, B. G. Dunavant, and G. K. Davis. Effect of dietary molybdenum and sulfate upon copper metabolism in sheep. J. Nutr. 99:177–183, 1969.
338. Marcilese, N. A., C. B. Ammerman, R. M. Valsecchi, B. G. Dunavant, and G. K. Davis. Effect of dietary molybdenum and sulfate upon urinary excretion of copper in sheep. J. Nutr. 100:1399–1406, 1970.
339. Margoshes, M., and B. L. Vallee. A cadmium protein from equine kidney cortex. J. Amer. Chem. Soc. 79:4813–4814, 1957. (letter)
340. Markowitz, H., G. E. Cartwright, and M. M. Wintrobe. Studies on copper metabolism. XXVII. The isolation and properties of an erythrocyte cuproprotein (erythrocuprein). J. Biol. Chem. 234:40–45, 1959.
341. Marsh, M. C., and R. K. Robinson. The treatment of fish-cultural waters for the removal of algae. Bull. Bur. Fish. 28(Part 2):871–890, 1908.
342. Marston, H. R. Cobalt, copper and molybdenum in the nutrition of animals and plants. Physiol. Rev. 32:66–121, 1952.
343. Mason, H. S. Preliminary remarks on polyphenoloxidase, pp. 339–341. In J. Peisach, P. Aisen and W. E. Blumberg, Eds. The Biochemistry of Copper. Proceedings of the Symposium on Copper in Biological Systems held at Arden House, Harriman, New York, September 8–10, 1965. New York: Academic Press, 1966.
344. Mattison, N. L. Enzymes of sphagnum moss, pp. 107–112. In Comprehensive Study of Physiologically Active Substances of Lower Plants. Leningrad: Akademiya Nauk S.S.S.R., Botanicheskie Institut, 1961. (in Russian)
345. Mayer, A. M. Ascorbic acid oxidase in germinating lettuce seeds and its inhibition. Physiol. Plant. 11:75–83, 1958.
346. McCabe, L. J. The problem of trace metals in water supply—An overview, pp. 1–9. In Proceedings. Sixteenth Water Quality Conference. Trace Metals in Water Supplies: Occurrence, Significance, and Control. Held Feb. 12–13, 1974 at University of Illinois at Urbana-Champaign.
347. McCabe, L. J., J. M. Symons, R. D. Lee, and G. G. Robeck. Survey of community water supply systems. J. Amer. Water Works Assoc. 62:670–687, 1970.
348. McCord, C. P. Metal fume fever as an immunological disease. Ind. Med. Surg. 29:101–106, 1960.
349. McCord, J. M., and I. Fridovich. Superoxide dismutase. An enzymic function for erythrocuprein (hemocuprein). J. Biol. Chem. 244:6049–6055, 1969.
350. McCosker, P. J. Observations on blood copper in the sheep. II. Chronic copper poisoning. Res. Vet. Sci. 9:103–116, 1968.
351. McEwen, C. M., Jr. Human plasma monoamine oxidase. I. Purification and identification. J. Biol. Chem. 240:2003–2010, 1965.
352. McEwen, C. M., Jr. Human plasma monoamine oxidase. II. Kinetic studies. J. Biol. Chem. 240:2011–2018, 1965.
353. McEwen, C. M., Jr., and D. C. Harrison. Abnormalities of serum monoamine oxidase in chronic congestive heart failure. J. Lab. Clin. Med. 65:546–559, 1965.
354. McIntyre, N., H. M. Clink, A. J. Levi, J. N. Cumings, and S. Sherlock. Hemolytic anemia in Wilson's disease. New Engl. J. Med. 276:439–444, 1967.
355. McMullen, W. Copper contamination of soft drinks from bottle pourers. Health Bull. (Edinburgh) 29:94–96, 1971.
356. McNatt, E. N., W. G. Campbell, Jr., and B. C. Callahan. Effects of dietary copper loading on livers of rats. I. Changes in subcellular acid phosphatases and detection

of an additional acid p-nitrophenylphosphatase in the cellular supernatant during copper loading. Amer. J. Path. 64:123–144, 1971.

357. Mehring, A. L., Jr., J. H. Brumbaugh, A. J. Sutherland, and H. W. Titus. The tolerance of growing chickens for dietary copper. Poult. Sci. 39:713–719, 1960.

358. Meister, A., and D. Wellner. Flavoprotein amino acid oxidases, pp. 609–648. In P. D. Boyer, H. Lardy and K. Myrbäck, Eds. The Enzymes. Vol. 7. Oxidation and Reduction (Part A), Nicotinamide Nucleotide-Linked Enzymes, Flavin Nucleotide-Linked Enzymes. (2nd ed.) New York: Academic Press, 1963.

359. Menkes, J. H., M. Alter, G. K. Steigleder, D. R. Weakley, and J. H. Sung. A sex-linked recessive disorder with retardation of growth, peculiar hair and focal cerebral and cerebellar degeneration. Pediatrics 29:764–779, 1962.

360. Metz, E. N., and A. L. Sagone, Jr. The effect of copper on the erythrocyte hexose monophosphate shunt pathway. J. Lab. Clin. Med. 80:405–413, 1972.

361. Milham, S., Jr., and T. Strong. Human arsenic exposure in relation to a copper smelter. Environ. Res. 7:176–182, 1974.

362. Milne, D. B., and P. H. Weswig. Effect of supplementary copper on blood and liver copper-containing fractions in rats. J. Nutr. 95:429–433, 1968.

363. Miltimore, J. E., and J. L. Mason. Copper to molybdenum ratio and molybdenum and copper concentrations in ruminant feeds. Can. J. Anim. Sci. 51:193–200, 1971.

364. Mischel, W. Die anorganischen Bestandteile der Placenta. VII. Der Kupfergehalt der reifen und unreifen, normalen und pathologischen menschlichen Placenta. Arch. Gynaek. 191:1–7, 1958.

365. Moffitt, A. E., Jr., and S. D. Murphy. Effect of excess and deficient copper intake on rat liver microsomal enzyme activity. Biochem. Pharmacol. 22:1463–1476, 1973.

366. Morell, A. G., R. A. Irvine, I. Sternlieb, I. H. Scheinberg, and G. Ashwell. Physical and chemical studies on ceruloplasmin. V. Metabolic studies on sialic acid-free ceruloplasmin in $vivo$. J. Biol. Chem. 243:155–159, 1968.

367. Morell, A. G., and I. H. Scheinberg. Heterogeneity of human ceruloplasmin. Science 131:930–932, 1960.

368. Morell, A. G., and I. H. Scheinberg. Preparation of an apoprotein from ceruloplasmin by reversible dissociation of copper. Science 127:588–590, 1958.

369. Morell, A. G., J. R. Shapiro, and I. H. Scheinberg. Copper binding protein of human liver, pp. 36–42. In J. M. Walshe and J. N. Cumings, Eds. Wilson's Disease. Some Current Concepts. Oxford: Blackwell Scientific Publications, Ltd., 1961.

370. Morell, A. G., J. Windsor, I. Sternlieb, and I. H. Scheinberg. Measurement of the concentration of ceruloplasmin in serum by determination of its oxidase activity, pp. 193–195. In F. W. Sunderman and F. W. Sunderman, Jr., Eds. Laboratory Diagnosis of Liver Diseases. St. Louis: Warren H. Green, Inc., 1968.

371. Morell, A. G., J. Windsor, I. Sternlieb, and I. H. Scheinberg. Spectrophotometric determination of microgram quantities of copper in biological materials, pp. 196–198. In F. W. Sunderman and F. W. Sunderman, Jr., Eds. Laboratory Diagnosis of Liver Diseases. St. Louis: Warren H. Green, Inc., 1968.

372. Morgan, J. M. Hepatic copper, manganese, and chromium content in bronchogenic carcinoma. Cancer 29:710–713, 1972.

373. Morrison, D. B., and T. P. Nash, Jr. The copper content of infant livers. J. Biol. Chem. 88:479–483, 1930.

374. Morrison, M., S. Horie, and H. S. Mason. Cytochrome c oxidase components.

II. A study of the copper in cytochrome c oxidase. J. Biol. Chem. 238:2220–2224, 1963.

375. Mortazavi, S. H., A. Bani-Hashemi, M. Mozafari, and A. Raffi. Value of serum copper measurement in lymphomas and several other malignancies. Cancer 29: 1193–1198, 1972.

376. Mosbach, R. Purification and some properties of laccase from *Polyporus versicolor*. Biochim. Biophys. Acta 73:204–212, 1963.

377. Mount, D. I. Chronic toxicity of copper to fathead minnows. (*Pimephales promelas*, rafinesque). Water Res. 2:215–223, 1968.

378. Mount, D. I., and C. E. Stephan. Chronic toxicity of copper to the fathead minnow (*Pimephales promelas*) in soft water. J. Fish. Res. Bd. Can. 26:2449–2457, 1969.

379. Murphy, L. S., and L. M. Walsh. Correction of micronutrient deficiencies with fertilizers, pp. 347–387. In J. J. Mortvedt, P. M. Giordano and W. L. Lindsay, Eds. Micronutrients in Agriculture. Proceedings of a Symposium, 1971. Madison, Wisc.: Soil Science Society of America, Inc., 1972.

380. Nair, P. M., and H. S. Mason. Reconstitution of cytochrome c oxidase from a copper-depleted enzyme and Cu^I. J. Biol. Chem. 242:1406–1415, 1967.

381. Nakamura, T. On the process of oxidation of hydroquinone by laccase, pp. 169–182. In M. S. Blois, Jr., H. W. Brown, R. M. Lemmon, R. O. Lindblom, and M. Weissbluth, Eds. Free Radicals in Biological Systems. Proceedings of a Symposium held at Stanford University, March 1960. New York: Academic Press, 1961.

382. Nakamura, T. Purification and physico-chemical properties of laccase. Biochim. Biophys. Acta 30:44–52, 1958.

383. Nakamura, T. Stoichiometric studies on the action of laccase. Biochim. Biophys. Acta 30:538–542, 1958.

384. Nakamura, T., N. Makino, and Y. Ogura. Purification and properties of ascorbate oxidase from cucumber. J. Biochem. (Tokyo) 64:189–195, 1968.

385. Nara, S., and K. T. Yasunobu. Some recent advances in the field of amine oxidases, pp. 423–436. In J. Peisach, P. Aisen and W. E. Blumberg, Eds. The Biochemistry of Copper. Proceedings of the Symposium on Copper in Biological Systems held at Arden House, Harriman, New York, September 8–10, 1965. New York: Academic Press, 1966.

386. National Research Council. Committee on Animal Nutrition. Nutrient Requirements of Swine. Nutrient Requirements of Domestic Animals No. 2 (7th rev. ed.). Washington, D.C.: National Academy of Sciences, 1973. 56 pp.

387. NCR-42 Committee on Swine Nutrition. Cooperative regional studies with growing swine: Effects of vitamin E and levels of supplementary copper during the growing-finishing period on gain, feed conversion and tissue copper storage in swine. J. Anim. Sci. 39:512–520, 1974.

388. Neilands, J. B., F. M. Strong, and C. A. Elvehjem. Molybdenum in the nutrition of the rat. J. Biol. Chem. 172:431–439, 1948.

389. Neumann, P. Z., and A. Sass-Kortsak. The state of copper in human serum: Evidence for an amino acid-bound fraction. J. Clin. Invest. 46:646–658, 1967.

390. Newbill, T. C., Jr. What the feed industry can expect from computers. Feed Stuffs 46(48):21, 22, 1974.

391. Newcomb, E. H. Dissociation of the effects of auxin on metabolism and growth of cultured tobacco pith. Physiol. Plant. 13:459–467, 1960.

392. Niedermeier, W., E. E. Creitz, and H. L. Holley. Trace metal composition of

synovial fluid from patients with rheumatoid arthritis. Arthr. Rheum. 5:439–444, 1962.

393. Nielsen, E. S., L. Kamp-Nielsen, and S. Wium-Andersen. The effect of deleterious concentrations of copper on the photosynthesis of *Chlorella pyrenoidosa*. Physiol. Plant. 22:1121–1133, 1969.

394. Nielsen, E. S., and S. Wium-Andersen. Copper ions as poison in the sea and in freshwater. Mar. Biol. 6:93–97, 1970.

395. Nostrand, I. F., and M. D. Glantz. Purification and properties of human liver monoamine oxidase. Arch. Biochem. Biophys. 158:1–11, 1973.

396. O'Dell, B. L., B. C. Hardwich, G. Reynolds, and J. E. Savage. Connective tissue defect in the chick resulting from copper deficiency. Proc. Soc. Exp. Biol. Med. 108:402–405, 1961.

397. O'Hara, J. Alterations in oxygen consumption by bluegills exposed to sublethal treatment with copper. Water Res. 5:321–327, 1971.

398. Ohta, K., Y. Okamoto, and O. Honda. Electron microscopic observations on the cerebrum of a case of Wilson's disease. Psychiatr. Neurol. Jap. 71:385–406, 1969. (in Japanese)

399. Okereke, T., I. Sternlieb, A. G. Morell, and I. H. Scheinberg. Systemic absorption of intrauterine copper. Science 177:358–360, 1972.

400. Omura, T. Studies on laccases of lacquer trees. I. Comparison of laccases obtained from *Rhus vernicifera* and *Rhus succedanea*. J. Biochem. (Toxyo) 50:264–272, 1961.

401. Omura, T. Studies on laccases of lacquer trees. III. Reconstruction of laccase from its protein and copper. J. Biochem. (Tokyo) 50:389–393, 1961.

402. O'Reilly, S., M. Pollycove, M. Tono, and L. Herradora. Abnormalities of the physiology of copper in Wilson's disease. II. The internal kinetics of copper. Arch. Neurol. 24:481–488, 1971.

403. O'Reilly, S., P. M. Weber, M. Oswald, and L. Shipley. Abnormalities of the physiology of copper in Wilson's disease. III. The excretion of copper. Arch. Neurol. 25:28–32, 1971.

404. Orlova, E. D. Effect of copper and molybdenum on the yield of spring wheat and the trace nutrient level in the grain, pp. 98–99. In V. R. Filippov, Ed. Trace Elements in the Biosphere and Their Utilization in Agriculture and Medicine of Siberia and the Far East, Reports of the Third Siberian Conference. Ulan-Ude, U.S.S.R.: Akademiya Nauk S.S.S.R. Sibirskoi Otdelenie Buryat Filial, 1971. (in Russian)

405. Osaki, S., D. A. Johnson, and E. Frieden. The possible significance of the ferrous oxidase activity of ceruloplasmin in normal human serum. J. Biol. Chem. 241: 2746–2751, 1966.

406. Osborn, S. B., C. N. Roberts, and J. M. Walshe. Uptake of radiocopper by the liver. A study of patients with Wilson's disease and various control groups. Clin. Sci. 24:13–22, 1963.

407. Osborn, S. B., and J. M. Walshe. Studies with radioactive copper (^{64}Cu and ^{67}Cu) in relation to the natural history of Wilson's disease. Lancet 1:346–350, 1967.

408. Osol, A., and R. Pratt, Eds. Cupric sulfate, pp. 356–357. In The United States Dispensatory. (27th ed.). Philadelphia: J. B. Lippincott, 1973.

409. Owen, C. A., Jr., and J. B. Hazelrig. Copper deficiency and copper toxicity in the rat. Amer. J. Physiol. 215:334–338, 1968.

410. Paine, C. H. Food-poisoning due to copper. Lancet 2:520, 1968.

411. Palmer, R. D., and W. K. Potter. The metabolism of nut grass (*Cyperus rotundus*

L.). IV. The activities of certain enzymes from tubers treated with amitrol. Weeds 7:511–517, 1959.

412. Pankratova, E. M. Increase in the physiological activity and crop yield of fruit-bearing trees by foliar nutrition. Translation Plant Physiol. (Fiziol. Rasten.) 7:479–483, 1961.

413. Parshikov, V. M. Boric nutrition as a factor affecting metabolism in hops. Akad. Nauk URSR Kiev Dopovidi 1958:338–342. (in Russian)

414. Passouant, P., J. Mirouze, P. Mary, M. Baldy-Moulinier, and P. Mahini. Epilepsie de type psycho-moteur manifestation évolutive d'un cas maladie de Wilson. J. Méd. Montpellier 4:147–149, 1969.

415. Passwell, J., B. E. Cohen, I. Ben Bassat, B. Ramot, M. Shchory, and U. Lavi. Hemolysis in Wilson's disease. The role of glucose-6-phosphate dehydrogenase inhibition. Israel J. Med. Sci. 6:549–554, 1970.

416. Patterson, J. B. E. Metal toxicities arising from industry, pp. 193–207. In Trace Elements in Soils and Crops. Ministry of Agriculture, Fisheries and Food Technical Bulletin 21. London: Her Majesty's Stationery Office, 1971.

417. Paulini, E. Copper molluscicides: Research and goals, pp. 155–170. In T. C. Cheng, Ed. Molluscicides in Schistosomiasis Control. New York: Academic Press, Inc., 1974.

418. Pedrero, E., and F. L. Kozelka. Effect of various pathological conditions on the copper content of human tissues. A.M.A. Arch. Path. 52:447–454, 1951.

419. Peisach, J., and W. G. Levine. A comparison of the enzymic activities of pig ceruloplasmin and Rhus vernicifera laccase. J. Biol. Chem. 240:2284–2289, 1965.

420. Peisach, J., W. G. Levine, and W. E. Blumberg. Structural properties of stellacyanin, a copper mucoprotein from Rhus vernicifera, the Japanese Lac Tree. J. Biol. Chem. 242:2847–2858, 1967.

421. Penton, Z. G., and C. R. Dawson. On the apoenzyme of ascorbate oxidase, pp. 222–239. In T. E. King, H. S. Mason, and M. Morrison, Eds. Oxidases and Related Redox Systems. Proceedings of a Symposium held in Amherst, Massachusetts, July 15–19, 1964. Vol. 1. New York: John Wiley & Sons, 1965.

422. Peterburgskii, A. V., Z. G. Antonova, and B. Nikolov. Physiological role of copper and molybdenum in the development of leguminous crops, pp. 40–57, In Ya. V. Pieve, Ed. Symposium on the Biological Role of Molybdenum, 1968. Moscow: Nauka, 1972. (in Russian)

423. Peters, R., M. Shorthouse, and J. M. Walshe. Studies on the toxicity of copper. II. The behaviour of microsomal membrane ATPase of the pigeon's brain tissue to copper and some other metallic substances. Proc. Roy. Soc. London B 166:285–294, 1966.

424. Peters, R., and J. M. Walshe. Studies on the toxicity of copper. I. The toxic action of copper in vivo and in vitro. Proc. Roy. Soc. London B 166:273–284, 1966.

425. Peters, R. A., M. Shorthouse, and J. M. Walshe. The effect of Cu^{2+} on the membrane ATPase and its relation to the initiation of convulsions. J. Physiol. 181: 27P–28P, 1965.

426. Peterson, R. E., and M. D. Bollier. Spectrophotometric determination of serum copper with biscyclohexanoneoxalyldihydrazone. Anal. Chem. 27:1195–1197, 1955.

427. Pickering, Q. H., and C. Henderson. The acute toxicity of some heavy metals to different species of warm water fishes. Air Water Pollut. 10:453–463, 1966.

428. Pierson, R. E., and W. A. Aanes. Treatment of chronic copper poisoning in sheep. J. Amer. Vet. Med. Assoc. 133:307–311, 1958.
429. Pimentel, J. C., and F. Marques. "Vineyard sprayer's lung": A new occupational disease. Thorax 24:678–688, 1969.
430. Pimental, J. C., and A. P. Menezes. Liver granulomas containing copper in vineyard sprayer's lung. A new etiology of hepatic granulomatosis. Amer. Rev. Respir. Dis. 111:189–195, 1975.
431. Pippy, J. H., and G. M. Hare. Relationship of river pollution to bacterial infection in salmon (*Salmo salar*) and suckers (*Catostomus commersoni*). Trans. Amer. Fish. Soc. 98:685–690, 1969.
432. Pojerová, A., and J. Továrek. Ceruloplasmin in early childhood. Acta Paediatr. 49:113–120, 1960.
433. Porter, C. C., D. C. Titus, B. E. Sanders, and E. V. C. Smith. Oxidation of serotonin in the presence of ceruloplasmin. Science 126:1014–1015, 1957.
434. Porter, H. Copper proteins in brain and liver in normal subjects and in cases of Wilson's disease, pp. 23–28. In D. Bergsma, Ed. Wilson's Disease–Birth Defects Original Article Series. Vol. 4, No. 2, New York: National Foundation, March of Dimes, 1968.
435. Porter, H. Neonatal hepatic mitochondrocuprein. III. Solubilization of the copper and protein from mitochondria of newborn liver by reduction with mercaptoethanol. Biochim. Biophys. Acta 154:236–238, 1968.
436. Porter, H. The cystine-rich copper storage protein of newborn liver and its possible relation to Wilson's disease. In Abstracts of Proceedings of the Third International Symposium on Wilson's Disease, Paris, September 20–21, 1973.
437. Powers, E. B. The goldfish (*Carassius carassius*) as a test animal in the study of toxicity. Illinois Biol. Monographs 4:127–193, 1917.
438. Prát, S. Die Erblichkeit der Resistenz gegen Kupfer. Ber. Dtsch. Bot. Ges. 52:65–67, 1934.
439. Price, C. A., and J. W. Quigley. A method for determining quantitative zinc requirements for growth. Soil Sci. 101:11–16, 1966.
440. Price, N. O., W. N. Linkous, and R. W. Engel. Minor element content of forage plants and soils. J. Agric. Food Chem. 3:226–229, 1955.
441. Puget, K., and A. M. Michelson. Isolation of a new copper-containing superoxide dismutase bacteriocuprein. Biochem. Biophys. Res. Commun. 58:830–838, 1974.
442. Pullar, E. M. The toxicity of various copper compounds and mixtures for domesticated birds. Austral. Vet. J. 16:147–162, 203–213, 1940.
443. Ragen, J. A., S. Nacht, G. R. Lee, C. R. Bishop, and G. E. Cartwright. Effect of ceruloplasmin on plasma iron in copper-deficient swine. Amer. J. Physiol. 217:1320–1323, 1969.
444. Rakhmanov, R. R., F. K. Sharafutdinova, and O. Dzhuraev. Effect of copper on certain physiologic and biochemical indices and productivity of cotton. Uzb. Biol. Zh. 17(3):26–27, 1973. (in Russian)
445. Raymont, J. E. G., and J. Shields. Toxicity of copper and chromium in the marine environment. Air Water Pollut. 7:435–443, 1963.
446. Recalde, L., and A. C. Blesa. Contribución al estudio del crecimiento en secciónes aisladas de coleoptilo de avena. VII. Intervención de algunos inhibidores del sistema ascorbico oxidasa en el crecimiento. Anales Edafologia Agrobiol. 20:379–386, 1961.
447. Rechenberger, J. Über die Harnkupferausscheidung bei der Nephrose. Dtsch. Z. Verdau. Stoffwechselkr. 17:199–205, 1957.

448. Reed, D. W., P. G. Passon, and D. E. Hultquist. Purification and properties of a pink copper protein from human erythrocytes. J. Biol. Chem. 245:2954-2961, 1970.

449. Reinhammar, B. Purification and properties of laccase and stellacyanin from *Rhus vernicifera*. Biochim. Biophys. Acta 205:35-47, 1970.

450. Reuther, W., and C. Labanauskas. Copper, pp. 157-179. In H. D. Chapman, Ed. Diagnostic Criteria for Plants and Soils. Berkeley: University of California, 1966.

451. Reuther, W., and P. F. Smith. Effects of high copper content of sandy soil on growth of citrus seedlings. Soil Sci. 75:219-224, 1953.

452. Reynolds, E. S., R. L. Tannen, and H. R. Tyler. The renal lesion in Wilson's disease. Amer. J. Med. 40:518-527, 1966.

453. Rice, E. W., R. E. Olson, and P. D. Sweeney. A study of serum copper and certain "acute-phase reactants" in alcoholics. Q. J. Stud. Alcohol 22:544-549, 1961.

454. Ritchie, H. D., R. W. Luecke, B. V. Baltzer, E. R. Miller, D. E. Ullrey, and J. A. Hoefer. Copper and zinc interrelationships in the pig. J. Nutr. 79:117-123, 1963.

455. Roche-Sicot, J., C. Sicot, G. Feldmann, B. Rueff, and J.-P. Benhamou. The syndrome of acute intravascular haemolysis and acute liver failure as a first manifestation of Wilson's disease. Digestion 8:447-448, 1973. (abstract)

456. Roeser, H. P., G. R. Lee, S. Nacht, and G. E. Cartwright. The role of ceruloplasmin in iron metabolism. J. Clin. Invest. 49:2408-2417, 1970.

457. Rosen, E. Copper within the eye. With the report of a case of typical sunflower cataract of the right eye and copper cataract involving the posterior capsule of the left eye. Amer. J. Ophthalmol. 32:248-252, 1949.

458. Ross, D. B. The diagnosis, prevention and treatment of chronic copper poisoning in housed lambs. Brit. Vet. J. 122:279-284, 1966.

459. Ross, D. B. The effect of oral ammonium molybdate and sodium sulfate given to lambs with high level copper concentrations. Res. Vet. Sci. 11:295-297, 1970.

460. Rydén, L. Studies on the Structure of Human Ceruloplasmin. Upsala: Acta Univ. Upsal. # 222, 1972. 35 pp.

461. St. George-Grambauer, T. D., and R. Rac. Hepatogenous chronic copper poisoning in sheep in South Australia due to the consumption of *Echium plantagineum* L. (Salvation Jane). Austral. Vet. J. 38:288-293, 1962.

462. Salmon, M. A., and T. Wright. Chronic copper poisoning presenting as pink disease. Arch. Dis. Child. 46:108-110, 1971.

463. Saltzer, E. I., and J. W. Wilson. Allergic contact dermatis due to copper. Arch. Derm. 98:375-376, 1968.

464. Sandberg, M., H. Gross, and O. M. Holly. Changes in retention of copper and iron in liver and spleen in chronic diseases accompanied by secondary anemia. Arch. Path. 33:834-844, 1942.

465. Sapozhnikova, E. V. Accumulation and conversion of ascorbic acid in Azerbaidzhan fruits and berries, pp. 80-86. In Trudy 1-oi (Pervoi) Vsesoyuznoi Konferentsii po Biologii Aktivnym Veshchestvam Plodov i Yagod, 1961. (in Russian)

466. Sarkar, B., and T. P. A. Kruck. Copper-amino acid complexes in human serum, pp. 183-196. In J. Peisach, P. Aisen and W. H. Blumberg, Eds. The Biochemistry of Copper. Proceedings of the Symposium on Copper in Biological Systems held at Arden House, Harriman, New York, September 8-10, 1965. New York: Academic Press, 1966.

467. Sass-Kortsak, A. Copper metabolism. Adv. Clin. Chem. 8:1-67, 1965.

468. Scharrer, K., and K. Mengel. Mikronährstoffe, pp. 456-518. In H. Linser, Ed. Handbuch der Pflanzenernährung und Düngung. Vol. I/1. New York: Springer-Verlag, 1969.

469. Scheinberg, I. H. Wilson's disease and copper-binding proteins. Science 185:1184, 1974.
470. Scheinberg, I. H., C. D. Cook, and J. A. Murphy. The concentration of copper and ceruloplasmin in maternal and infant plasma at delivery. J. Clin. Invest. 33:963, 1954. (abstract)
471. Scheinberg, I. H., and D. Gitlin. Deficiency of ceruloplasmin in patients with hepatolenticular degeneration (Wilson's disease). Science 116:484–485, 1952.
472. Scheinberg, I. H., and A. G. Morell. Ceruloplasmin, pp. 306–319. In G. L. Eichhorn, Ed. Inorganic Biochemistry. Vol. 1. Amsterdam: Elsevier Scientific Publishing Company, 1973.
473. Scheinberg, I. H., and I. Sternlieb. Copper metabolism. Pharmacol. Rev. 12:355–381, 1960.
474. Scheinberg, I. H., and I. Sternlieb. Metabolism of trace metals, pp. 1321–1334. In P. K. Bondy, Ed. Duncan's Diseases of Metabolism. Vol. 2. Endocrinology and Nutrition. (6th ed.) Philadelphia: W. B. Saunders Co., 1969.
475. Scheinberg, I. H., and I. Sternlieb. The liver in Wilson's disease. Gastroenterology 37:550–564, 1959.
476. Scheinberg, I. H., and I. Sternlieb. Wilson's disease. Ann. Rev. Med. 16:119–134, 1965.
477. Scheinberg, I. H., and I. Sternlieb. Wilson's disease, pp. 247–264. In G. E. Gaull, Ed. Biology of Brain Dysfunction. Vol. 3. New York: Plenum Press, 1975.
478. Schiötz, E. H. Metal fever produced by copper dust, pp. 798–801. In The Proceedings of the Ninth International Congress on Industrial Medicine, London, September 13–17, 1948. Bristol: John Wright and Sons Ltd., 1949.
479. Schroeder, H. A. A sensible look at air pollution by metals. Arch. Environ. Health 21:798–806, 1970.
480. Schroeder, H. A., A. P. Nason, I. H. Tipton, and J. J. Balassa. Essential trace metals in man. Copper. J. Chron. Dis. 19:1007–1034, 1966.
481. Schulman, S. Wilson's disease, pp. 1139–1152. In J. Minckler, Ed. Pathology of the Nervous System, Vol. 1. New York: McGraw-Hill Book Co., 1968.
482. Schutz, G., and P. Feigelson. Purification and properties of rat liver tryptophan oxygenase. J. Biol. Chem. 247-5327–5332, 1972.
483. Seely, J. R., G. B. Humphrey, and B. J. Matter. Copper deficiency in a premature infant fed an iron-fortified formula. Clin. Res. 20:107, 1972. (abstract)
484. Seely, J. R., G. B. Humphrey, and B. J. Matter. Copper deficiency in a premature infant fed an iron-fortified formula. New Engl. J. Med. 286:109–110, 1972. (letter)
485. Semple, A. B., W. H. Parry, and D. E. Phillips. Acute copper poisoning. An outbreak traced to contaminated water from a corroded geyser. Lancet 2:700–701, 1960.
486. Serova, Z. Y. The metabolic activity of plants attacked by rust fungi. Akad. Nauk Belorussk. SSR Dokl. 5:405–408, 1961. (in Russian)
487. Shields, G. S., H. Markowitz, W. H. Klassen, G. E. Cartwright, and M. M. Wintrobe. Studies on copper metabolism. XXXI. Erythrocyte copper. J. Clin. Invest. 40:2007–2015, 1961.
488. Shishkanu, G. V., and N. V. Semenova. Effect of trace elements on the photosynthetic apparatus of vegetatively reproduced apple tree stock, pp. 38–71. In B. L. Dorokhov, Ed. Photosynthesis and Pigments of the Basic Agricultural Plants in Moldavia. Kishinev, U.S.S.R.: Tzvestiya Tsentralata Kommunistii Partii Moldavii, 1970. (in Russian).
489. Shokeir, M. H. K., and D. C. Shreffler. Two new ceruloplasmin variants in Negroes—data on three populations. Biochem. Genet. 4:517–528, 1970.

490. Shreffler, D. C., G. J. Brewer, J. C. Gall, and M. S. Honeyman. Electrophoretic variation in human serum ceruloplasmin: A new genetic polymorphism. Biochem. Genet. 1:101–115, 1967.

491. Siegel, R. C., S. R. Pinnell, and G. R. Martin. Cross-linking of collagen and elastin. Properties of lysyl oxidase. Biochemistry 9:4486–4492, 1970.

492. Sinha, S. N., and E. R. Gabrieli. Serum copper and zinc levels in various pathologic conditions. Amer. J. Clin. Path. 54:570–577, 1970.

493. Slinger, S. J., W. F. Pepper, and I. R. Sibbald. Copper sulphate and penicillin as supplements for chicks. Poult. Sci. 41:341–342, 1962.

494. Smallwood, R. A., H. A. Williams, V. M. Rosenoer, and S. Sherlock. Liver-copper levels in lever disease. Studies using neutron activation analysis. Lancet 2:1310–1313, 1968.

495. Smith, C. K., and R. H. Mattson. Seizures in Wilson's disease. Neurology 17:1121–1123, 1967.

496. Smith, H. The distribution of antimony, arsenic, copper and zinc in human tissue. J. Forensic Sci. Soc. 7:97–102, 1967.

497. Smith, H. M. Effects of sulfhydryl blockage on axonal function. J. Cell. Comp. Physiol. 51:161–171, 1958.

498. Smith, M. S. Responses of chicks to dietary supplements of copper sulphate. Brit. Poult. Sci. 10:97–108, 1969.

499. Smith, S. E., and E. J. Larson. Zinc toxicity in rats; antagonistic effects of copper and liver. J. Biol. Chem. 163:29–38, 1946.

500. Sommer, A. L. Copper as an essential element for plant growth. Plant Physiol. 6:339–345, 1931.

501. Sourkes, T. L. Influence of specific nutrients on catecholamine synthesis and metabolism. Pharmacol. Rev. 24:349–359, 1972.

502. Sprague, J. B. Promising anti-pollutant: Chelating agent NTA protects fish from copper and zinc. Nature 220:1345–1346, 1968.

503. Sprague, J. B., P. F. Elson, and R. L. Saunders. Sublethal copper-zinc pollution in a salmon river—a field and laboratory study. Adv. Water Pollut. Res. 1:61–82, 1964.

504. Spray, C. M., and E. M. Widdowson. The effect of growth and development on the composition of mammals. Brit. J. Nutr. 4:332–353, 1950.

505. Stahr, H. M., Ed. Analytical Toxicology Methods Manual. Ames: I.S.U. (Iowa State University) Foundation, 1976. [300 pp.]

506. Stansell, M. J., and H. F. Deutsch. Physicochemical studies of crystalline human erythrocuprein. J. Biol. Chem. 240:4306–4311, 1965.

507. Starcher, B. C. Studies on the mechanism of copper absorption in the chick. J. Nutr. 97:321–326, 1969.

508. Stark, G. R., and C. R. Dawson. Ascorbic acid oxidase, pp. 297–311. In P. D. Boyer, H. Lardy and K. Myrbäck, Eds. The Enzymes. Vol. 8. Oxidation and Reduction (Part B), Metal-Porphyrin Enzymes, Other Oxidases, Oxygenation, Topical Subject Index: Vol. 1-8. (2nd ed.) New York: Academic Press, 1963.

509. Stark, G. R., and C. R. Dawson. On the accessibility of sulfhydryl groups in ascorbic acid oxidase. J. Biol. Chem. 237:712–716, 1962.

510. Stein, R. S., D. Jenkins, and M. Korns. Death after use of cupric sulfate as emetic. J.A.M.A. 235:801, 1976. (letter)

511. Sternlieb, I. Evolution of the hepatic lesion in Wilson's disease (hepatolenticular degeneration). Prog. Liver Dis. 4:511–525, 1972.

512. Sternlieb, I. Gastrointestinal copper absorption in man. Gastroenterology 52: 1038–1041, 1967.

513. Sternlieb, I. The beneficial and adverse effects of penicillamine, pp. 183–190. In H. Popper and K. Becker, Eds. Collagen Metabolism and the Liver. Proceedings of an International Conference held in Freiburg i. Br., W. Germany, October 10–11, 1973. New York: Stratton Intercontinental Medical Book Corporation, 1975.

514. Sternlieb, I. The Kayser-Fleischer ring. Med. Radiogr. Photogr. 42:14–15, 1966.

515. Sternlieb, I., R. C. Harris, and I. H. Scheinberg. Le Cuivre dans la cirrhose biliaire de l'enfant. Rev. Int. Hépatol. 16:1105–1110, 1966.

516. Sternlieb, I., and H. D. Janowitz. Absorption of copper in malabsorption syndromes. J. Clin. Invest. 43:1049–1055, 1964.

517. Sternlieb, I., A. G. Morell, and I. H. Scheinberg. The uniqueness of ceruloplasmin in the study of plasma protein synthesis. Trans. Assoc. Amer. Physicians 75:228–234, 1962.

518. Sternlieb, I., A. G. Morell, W. D. Tucker, M. W. Greene, and I. H. Scheinberg. The incorporation of copper into ceruloplasmin in vivo: Studies with copper[64] and copper[67]. J. Clin. Invest. 40:1834–1840, 1961.

519. Sternlieb, I., J. I. Sandson, A. G. Morell, E. Korotkin, and I. H. Scheinberg. Nonceruloplasmin copper in rheumatoid arthritis. Arthr. Rheum. 12:458–459, 1969. (letter)

520. Sternlieb, I., and I. H. Scheinberg. Ceruloplasmin in health and disease. Ann. N.Y. Acad. Sci. 94:71–76, 1961.

521. Sternlieb, I., and I. H. Scheinberg. Penicillamine therapy in hepatolenticular degeneration. J.A.M.A. 189:748–754, 1964.

522. Sternlieb, I., and I. H. Scheinberg. Prevention of Wilson's disease in asymptomatic patients. New Engl. J. Med. 278:352–359, 1968.

523. Sternlieb, I., and I. H. Scheinberg. Radiocopper in diagnosing liver disease. Semin. Nucl. Med. 2:176–188, 1972.

524. Sternlieb, I., and I. H. Scheinberg. Wilson's disease, pp. 328–336. In F. Schaffner, S. Sherlock, and C. M. Leevy, Eds. The Liver and Its Diseases. New York: Intercontinental Medical Book Crop., 1974.

525. Sternlieb, I., I. H. Scheinberg, and J. M. Walshe. Bleeding oesophageal varices in patients with Wilson's disease. Lancet 1:638–641, 1970.

526. Sternlieb, I., C. J. A. van den Hamer, A. G. Morell, S. Alpert, G. Gregoriadis, and I. H. Scheinberg. Lysosomal defect of hepatic copper excretion in Wilson's disease (hepatolenticular degeneration). Gastroenterology 64:99–105, 1973.

527. Steyn-Parvé, E. P., and H. Beinert. On the mechanism of dehydrogenase of fatty acyl derivatives of coenzyme A. VII. The nature of the green color of butyryl dehydrogenase. J. Biol. Chem. 233:853–861, 1958.

528. Stocks, P., and R. I. Davies. Zinc and copper content of soils associated with the incidence of cancer of the stomach and other organs. Brit. J. Cancer 18:14–24, 1964.

528a. Stokinger, H. D. Copper, pp. 1033–1037. In D. W. Fassett and D. A. Irish, Eds. Toxicology. Vol. 2. In F. A. Patty, Ed. Industrial Hygiene and Toxicology. (2 Vol.) New York: Interscience Publishers, 1963.

529. Stout, P. R. Micronutrient needs for plant growth, pp. 21–23. In Proceedings. Ninth Annual California Fertilizer Conference, March 27 and 28, 1961, California State Polytechnic College, Pomona, California.

530. Strickland, G. T., W. M. Beckner, and M.-L. Leu. Absorption of copper in homozygotes and heterozygotes for Wilson's disease and controls: Isotope tracer studies with [67]Cu and [64]Cu. Clin. Sci. 43:617–625, 1972.

531. Strickland, G. T., W. M. Beckner, M.-L. Leu, and S. O'Reilly. Copper-67 studies in Wilson's disease patients and their families. Clin. Res. 17:396, 1969. (abstract)

532. Strothkamp, K. G., and C. R. Dawson. Concerning the quaternary structure of ascorbate oxidase. Biochemistry 13:434–440, 1974.

533. Sturua, L. I. Effect of the presowing treatment of spring wheat seeds on its sowing qualities and yield. Dokl. TSKhA 180(2):11–14, 1972. (in Russian)

534. Summerlin, W. T., A. I. Walder, and J. A. Moncrief. White phosphorus burns and massive hemolysis. J. Trauma 7:476–484, 1967.

535. Sunthankar, S. V., and C. R. Dawson. The structural identification of the olefinic components of Japanese lac urushiol. J. Amer. Chem. Soc. 76:5070–5074, 1954.

536. Sussman, W., and I. H. Scheinberg. Disappearance of Kayser-Fleischer rings. Effects of penicillamine. Arch. Ophthalmol. 82:738–741, 1969.

537. Sutter, M. D., D. C. Rawson, J. A. McKeown, and A. R. Haskell. Chronic copper toxicosis in sheep. Amer. J. Vet. Res. 19:890–892, 1958.

538. Suttle, N. F., K. W. Angus, D. I. Nisbet, and A. C. Field. Osteoporosis in copper-depleted lambs. J. Comp. Path. 82:93–97, 1972.

539. Suttle, N. F., and C. F. Mills. Studies of the toxicity of copper to pigs. 1. Effects of oral supplements of zinc and iron salts on the development of copper toxicosis. Brit. J. Nutr. 20:135–148, 1966.

540. Suttle, N. F., and C. F. Mills. Studies of the toxicity of copper to pigs. 2. Effect of protein source and other dietary components on the response to high and moderate intakes of copper. Brit. J. Nutr. 20:149–161, 1966.

541. Suveges, T., F. Ratz, and G. Salyi. Pathogenesis of chronic copper poisoning in lambs. Acta Vet. (Budapest) 21:383–391, 1971.

542. Swaine, D. J. The Trace-Element Content of Soils. Technical Communication No. 48 of the Commonwealth Bureau of Soil Science, Rothamsted Experimental Station, Harpenden. Farnham Royal Bucks, England: Commonwealth Agricultural Bureaux, 1955. 167 pp.

543. Swaine, D. J., and R. L. Mitchell. Trace element distribution in soil profiles. J. Soil Sci. 11:347–368, 1960.

544. Szent-Györgyi, A. Observations on the function of peroxidase systems and the chemistry of the adrenal cortex. Description of a new carbohydrate derivative. Biochem. J. 22:1387–1409, 1928.

545. Tani, P., and K. Kokkola. Serum iron, copper, and iron-binding capacity in brochogenic pulmonary carcinoma. Scand. J. Resp. Dis. Suppl. 80:121–128, 1972.

546. Tatum, H. J. Metallic copper as an intrauterine contraceptive agent. Amer. J. Obstet. Gynecol. 117:602–618, 1973.

547. Taylor, M., and S. Thomke. Effect of high-level copper on the depot fat of bacon pigs. Nature 201:1246, 1964.

548. Tessmer, C. F., M. Hrgovcic, B. W. Brown, J. Wilbur, and F. B. Thomas. Serum copper correlations with bone marrow. Cancer 29:173–179, 1972.

549. Tessmer, C. F., M. Hrgovcic, F. B. Thomas, L. M. Fuller, and J. R. Castro. Serum copper as an index of tumor response to radiotherapy. Radiology 106:635–639, 1973.

550. Tessmer, C. F., M. Hrgovcic, F. B. Thomas, J. Wilbur, and D. M. Mumford. Long-term serum copper studies in acute leukemia in children. Cancer 30:358–365, 1972.

551. Thornton, I., W. J. Atkinson, J. S. Webb, and D. B. R. Poole. Geochemical reconnaissance and bovine hypocuprosis in Co. Limerick, Ireland. Irish J. Agric. Res. 5:280–283, 1966.

552. Thornton, I., G. F. Kershaw, and M. G. Davies. An investigation into sub-clinical copper deficiency in cattle. Vet. Rec. 90:11–12, 1972.

553. Tiffin, L. O. Translocation of micronutrients in plants, pp. 199–229. In J. J. Mortvedt, P. M. Giordano and W. L. Lindsay, Eds. Micronutrients in Agriculture. Proceedings of a Symposium, 1971. Madison, Wisc.: Soil Science Society of America, Inc., 1972.

554. Ting-Beall, H. P., D. A. Clark, C. H. Suelter, and W. W. Wells. Studies on the interaction of chick brain microsomal ($Na^+ + K^+$)-ATPase with copper. Biochim. Biophys. Acta 291:229–236, 1973.

555. Tipton, I. H., P. L. Stewart, and P. G. Martin. Trace elements in diets and excreta. Health Phys. 12:1683–1689, 1966.

556. Tisdale, S. L., and W. L. Nelson. Soil Fertility and Fertilizers. (3rd ed.) New York: MacMillan Publishing Co., Inc., 1975. 694 pp.

557. Titcomb, J. W. The use of copper sulphate for the destruction of obnoxious fishes in lakes and ponds. Trans. Amer. Fish. Soc. 44:20–26, 1914.

558. Todd, J. R. Chronic copper poisoning in farm animals. Vet. Bull. 32:573–580, 1962.

559. Todd, J. R. Chronic copper toxicity of ruminants. Proc. Nutr. Soc. 28:189–197, 1969.

560. Todd, J. R. Copper, molybdenum and sulphur contents of oats and barley in relation to chronic copper poisoning in housed sheep. J. Agric. Sci. 79:191–195, 1972.

561. Todd, J. R., J. F. Gracey, and R. H. Thompson. Studies on chronic copper poisoning. I. Toxicity of copper sulphate and copper acetate in sheep. Brit. Vet. J. 118: 482–491, 1962.

562. Todd, J. R., and R. H. Thompson. Studies on chronic copper poisoning. II. Biochemical studies on the blood of sheep during the haemolytic crisis. Brit. Vet. J. 119:161–173, 1963.

563. Todd, J. R., and R. H. Thompson. Studies on chronic copper poisoning. III. Effects of copper acetate injected into the blood stream of sheep. J. Comp. Path. Therap. 74:542–551, 1964.

564. Todd, J. R., and R. H. Thompson. Studies on chronic copper poisoning. IV. Biochemistry of the toxic syndrome in the calf. Brit. Vet. J. 121:90–97, 1965.

565. Tokudome, S., and M. Kuratsune. A cohort study on mortality from cancer and other causes among workers at a metal refinery. Int. J. Cancer 17:310–317, 1976.

566. Tönz, O., H. U. Furrer, and U. Bangerter. Kupferinduzierte Hämolyse bei Morbus Wilson. Schweiz. Med. Wschr. 101:1800–1802, 1971.

567. Topham, R. W., and E. Frieden. Identification and purification of a non-ceruloplasmin ferroxidase of human serum. J. Biol. Chem. 245:6698–6705, 1970.

568. Trachtenberg, D. I. Allergic response to copper—its possible gingival implications. J. Periodontol. 43:705–707, 1972.

569. Tu, J.-B., R. Q. Blackwell, and T.-Y. Hou. Tissue copper levels in Chinese patients with Wilson's disease. Neurology 13:155–159, 1963.

570. Tyson, T. L., H. H. Holmes, and C. Ragan. Copper therapy of rheumatoid arthritis. Amer. J. Med. Sci. 220:418–420, 1950.

571. Underwood, E. J. Copper, pp. 57–115. In Trace Elements in Human and Animal Nutrition. (3rd ed.) New York: Academic Press, 1971.

572. Underwood, E. J. Trace Elements in Human and Animal Nutrition. (3rd ed.) New York: Academic Press, 1971. 543 pp.

573. Underwood, P. C., J. H. Collins, C. G. Durbin, F. A. Hodges, and H. E. Zimmerman, Jr. Critical tests with copper sulfate for experimental moniliasis (crop mycosis) of chickens and turkeys. Poult. Sci. 35:599–605, 1956.

574. U.S. Bureau of Mines. Copper, pp. 46–47. In Commodity Data Summaries, 1976.

An Up-to-Date Summary of 95 Mineral Commodities. Washington, D.C.: U.S. Department of the Interior, 1976.

575.   U.S. Department of Health, Education, and Welfare. Food and Drug Administration. Restriction on level of copper in animal feed. Fed. Reg. 38:25694–25696, 1973.

576.   U.S. Department of Health, Education, and Welfare. National Communicable Disease Center. Foodborne Outbreaks. Annual Summary 1968. 40 pp.

577.   U.S. Department of Health, Education, and Welfare. Public Health Service. National Communicable Disease Center. Foodborne Outbreaks. Annual Summary 1969. 35 pp.

578.   U.S. Department of Health, Education, and Welfare. Public Health Service Center for Disease Control. Foodborne Outbreaks. Annual Summary 1972. 45 pp.

579.   U.S. Department of Health, Education, and Welfare. Public Health Service Center for Disease Control. Foodborne Outbreaks. January–June 1971. DHEW Publ. No. (HSM) 72-8135. Washington, D.C.: U.S. Government Printing Office, 1972. 30 pp.

580.   U.S. Department of Health, Education, and Welfare, Public Health Service, Environmental Health Service, Bureau of Water Hygiene. Community Water Supply Study: Significance of National Findings. Washington, D.C.: U.S. Department of Health, Education, and Welfare, 1970. 13 pp.

581.   U.S. Department of Health, Education, and Welfare. Public Health Service. National Air Pollution Control Administration. Air Quality Data from the National Air Surveillance Networks and Contribution State and Local Networks. 1966 Edition. NAPCA Publ. APTD 68-9. Washington, D.C.: U.S. Government Printing Office, 1968. 157 pp.

582.   U.S. Environmental Protection Agency. Water Quality Office. Analytical Quality Control Laboratories. Methods for Chemical Analysis of Water and Wastes 1971. Washington, D.C.: U.S. Government Printing Office, 1971. 312 pp.

583.   Vallee, B. L., and W. E. C. Wacker. Metalloproteins. In H. Neurath, Ed. The Proteins. (2nd ed.) Vol. 5. New York: Academic Press, 1970. 192 pp.

584.   van Adrichem, P. W. M. Changes in the activity of serum enzymes and in the lactic dehydrogenase isoenzyme pattern in chronic copper intoxication in sheep. Tijdschr. Diergeneesk. 90:1371–1381, 1965. (in Dutch)

585.   van den Hamer, C. J. A., A. G. Morell, and I. H. Scheinberg. A study of the copper content of β-mercaptopyruvate trans-sulfurase. J. Biol. Chem. 242:2514–2516, 1967.

586.   Van Reen, R. Effects of excessive dietary zinc in the rat and the interrelationship with copper. Arch. Biochem. Biophys. 46:337–344, 1953.

587.   Verity, M. A., and J. K. Gambell. Studies of copper ion-induced mitochondrial swelling *in vitro*. Biochem. J. 108:289–295, 1968.

588.   Verity, M. A., J. K. Gambell, A. R. Reith, and W. J. Brown. Subcellular distribution and enzyme changes following subacute copper intoxication. Lab. Invest. 16:580–590, 1967.

589.   Vernon, L. P., B. Ke, and E. R. Shaw. Relationship of P700, electron spin resonance signal, and photochemical activity of a small chloroplast particle obtained by the action of Triton X-100. Biochemistry 6:2210–2220, 1967.

590.   Villar, T. G. Vineyard sprayer's lung. Clinical aspects. Amer. Rev. Respir. Dis. 110:545–555, 1974.

591.   Vilter, R. W., R. C. Bozian, E. V. Hess, D. C. Zellner, and H. G. Petering. Manifestations of copper deficiency in a patient with systemic sclerosis of intravenous hyperalimentation. New Engl. J. Med. 291–188–191, 1974.

592. Vines, H. M., and M. F. Oberbacher. Ascorbic acid oxidase in citrus. Proc. Florida State Hort. Soc. 75:283–286, 1962.

593. Voelker, R. W., Jr., and W. W. Carlton. Effect of ascorbic acid on copper deficiency in miniature swine. Amer. J. Vet. Res. 30:1825–1830, 1969.

594. Vogel, F. S. Nephrotoxic properties of copper under experimental conditions in mice, with special reference to the pathogenesis of the renal alterations in Wilson's disease. Amer. J. Path. 36:699–711, 1960.

595. Vogel, F. S., and J. W. Evans. Morphologic alterations produced by copper in neural tissues with consideration of the role of the metal in the pathogenesis of Wilson's disease. J. Exp. Med. 113:997–1004, 1961.

596. Vogel, F. S., and L. Kemper. Biochemical reactions of copper within neural mitochondria, with consideration of the role of the metal in the pathogenesis of Wilson's disease. Lab. Invest. 12:171–179, 1963.

597. Vohra, P., and F. H. Kratzer. Zinc, copper and manganese toxicities in turkey poults and their alleviation by EDTA. Poult. Sci. 47:699–704, 1968.

598. Voinar, A. I., and V. N. Galakhova. Effect of copper on the glycogen and lipid contents of the liver. Ukr. Biokhim. Zh. 34:504–506, 1962. (in Russian)

599. Vrublevskaya, D. G. Changes in activity of oxidative enzymes and accumulation of protein in the developing plant. Uch. Zap. Tomsk. Gos. Univ. 44:208–214, 1962. (in Russian)

600. Wahal, P. K., V. P. Mittal, and O. P. Bansal. Renal complications in acute copper sulphate poisoning. Indian Pract. 18:807–812, 1965.

601. Waldmann, T. A., A. G. Morell, R. D. Wochner, W. Strober, and I. Sternlieb. Measurement of gastrointestinal protein loss using ceruloplasmin labeled with ^{67}copper. J. Clin. Invest. 46:10–20, 1967.

602. Wallace, H. D., J. T. McCall, B. Bass, and G. E. Combs. High level copper for growing-finishing swine. J. Anim. Sci. 19:1153–1163, 1960.

603. Wallace, T. The Diagnosis of Mineral Deficiencies in Plants by Visual Symptoms. A Colour Atlas and Guide. (2nd Amer. ed.) New York: Chemical Publishing Co., 1961. 283 pp.

604. Walshe, J. M. Copper chelation in patients with Wilson's disease. Q. J. Med. 42:441–452, 1973.

605. Walshe, J. M. Penicillamine, a new oral therapy for Wilson's disease. Amer. J. Med. 21:487–495, 1956.

606. Walshe, J. M. Triethylene tetramine. Lancet 2:154, 1970. (letter)

607. Walshe, J. M. Wilson's disease, a review, pp. 475–498. In J. Peisach, P. Aisen, and W. E. Blumberg, Eds. The Biochemistry of Copper. Proceedings of the Symposium on Copper in Biological Systems held at Arden House, Harriman, New York, September 8–10, 1965. New York: Academic Press, 1966.

608. Walshe, J. M., and J. Briggs. Caeruloplasmin in liver disease. A diagnostic pitfall. Lancet 2:263–265, 1962.

609. Warren, R. L., A. M. Jelliffe, J. V. Watson, and C. B. Hobbs. Prolonged observations on variations in the serum copper in Hodgkin's disease. Clin. Radiol. 20: 247–256, 1969.

610. Weast, R. C., Ed. [Table] Crystal ionic radii of the elements, p. F-198. In CRC Handbook of Chemistry and Physics. (55th ed.) Cleveland, Ohio: Chemical Rubber Company Press, 1974.

611. Wellborn, T. L., Jr. Toxicity of nine therapeutic and herbicidal compounds to striped bass. Prog. Fish Cult. 31:27–32, 1969.

612. Weser, U., G. Barth, C. Djerassi, H.-J. Hartmann, P. Krauss, G. Voelcker, W.

Voelter, and W. Vietsch. A study on purified apo-erythrocuprein. Biochim. Biophys. Acta 278:28–44, 1972.

613. Weser, U., H. Rupp, F. Donay, F. Linnemann, W. Voelter, W. Voetsch, and G. Jung. Characterization of Cd, Zn-thionein (metallothionein) isolated from rat and chicken liver. Eur. J. Biochem. 39:127–140, 1973.

614. Widdowson, E. M. Chemical composition of newly born mammals. Nature 166:626–628, 1950.

615. Wiederanders, R. E. Copper loading in the turkey. Proc. Soc. Exp. Biol. Med. 128:627–629, 1968.

616. Williams, D. M., G. R. Lee, and G. E. Cartwright. Mitochondrial iron metabolism. Fed. Proc. 32:924, 1973. (abstract)

617. Willms, B., K. G. Blume, and G. W. Löhr. Hämolytische Anämie bei Morbus Wilson (Hepatolentikuläre Degeneration). Klin. Wschr. 50:995–1002, 1972.

618. Wilson, J. E., and M. E. Lahey. Failure to induce dietary deficiency of copper in premature infants. Pediatrics 25:40–49, 1960.

619. Wilson, M. L., A. A. Iodice, M. P. Schulman, and D. A. Richert. Studies on liver δ-aminolevulinic acid dehydrase. Fed. Proc. 18:352, 1959. (abstract)

620. Wilson, S. A. K. Progressive lenticular degeneration: A familial nervous disease associated with cirrhosis of the liver. Brain 34:295–509, 1912.

621. Winge, D. R., R. Premakumar, R. D. Wiley, and K. V. Rajagopalan. Copper-chelatin: Purification and properties of a copper-binding protein from rat liver. Arch. Biochem. Biophys. 170:253–266, 1975.

622. Wiśniewski, H., M. Smialek, H. Szydlowska, and T. Zalewska. Quantitative topography of copper in Wilson's disease and in porto-systemic encephalopathy. Neuropatol. Pol. 5:92–103, 1967.

623. Wolff, S. M. Copper deposition in the rat. A.M.A. Arch. Path. 69:217–223, 1960.

624. Wolff, S. M. Renal lesions in Wilson's disease. Lancet 1:843–845, 1964.

625. Woodside, J. D. Copper poisoning from pig slurry–is there a risk? Agric. North. Ireland 48:52–56, 1973.

626. Worwood, M., and D. M. Taylor. Subcellular distribution of copper in rat liver after biliary obstruction. Biochem. Med. 3:105–116, 1969.

627. Worwood, M., D. M. Taylor, and A. H. Hunt. Copper and manganese concentrations in biliary cirrhosis of liver. Brit. Med. J. 3:344–346, 1968.

628. Yamada, H., and K. T. Yasunobu. Monoamine oxidase. II. Copper, one of the prosthetic groups of plasma monoamine oxidase. J. Biol. Chem. 237:3077–3082, 1962.

629. Yamazaki, I. The reduction of cytochrome c by enzyme-generated ascorbic free radical. J. Biol. Chem. 237:224–229, 1962.

630. Yamazaki, I., and L. H. Piette. Mechanism of free radical formation and disappearance during the ascorbic acid oxidase and peroxidase reactions. Biochim. Biophys. Acta 50:62–69, 1961.

631. Yong, F. C., and T. E. King. Studies on cytochrome oxidase. IX. Hemecopper interaction. J. Biol. Chem. 247:6384–6388, 1972.

632. Yoshida, H. Chemistry of lacquer (Urushi). Part I. J. Chem. Soc. Trans. 43:472–486, 1883.

633. Zadorojny, G. P., and I. N. Venichenko. Application of copper salts as fertilizers under cotton plant. Agrokhimiya 1973(9):109–118. (in Russian)

634. Zeller, E. A. Diamine Oxidases, pp. 313–335. In P. D. Boyer, H. Lardy, and K. Myrbäck, Eds. The Enzymes. Vol. 8. Oxidation and Reduction (Part B), Metal-Prophyrin Enzymes, Other Oxidases, Oxygenation, Topical Subject Index: Volumes 1–8. (2nd ed.) New York: Academic Press, 1963.

635. Zeller, E. A. Monoamine and polyamine analogues, pp. 53–78. In R. M. Hochster and J. H. Quastel, Eds. Metabolic Inhibitors. A Comprehensive Treatise. Vol. 2. New York: Academic Press, 1963.

636. Zeller, E. A., and J. R. Fouts. Enzymes as primary targets of drugs. Ann. Rev. Pharmacol. 3:9–32, 1963.

637. Zipper, J. A., H. J. Tatum, L. Pastene, M. Medel, and M. Rivera. Metallic copper as an intrauterine contraceptive adjunct to the "T" device. Amer. J. Obstet. Gynec. 105:1274–1278,1968.

638. Zureck, I. Über das Auftreten von Kupfervergiftungen in Geflügelbeständen. Monatsh. Veterinärmed. 26:458–460, 1971.

Index